abc's of Antennas

by

Allan Lytel and Farl J. Waters

HOWARD W. SAMS & CO., INC.
THE BOBBS-MERRILL CO., INC.
INDIANAPOLIS · KANSAS CITY · NEW YORK

SECOND EDITION

FIRST PRINTING—1976

International Standard Book Number: 0-672-21300-1
Library of Congress Catalog Card Number: 76-19695

Preface

The most important link in the entire chain of radio-wave transmission and reception is the antenna. Without a good antenna, communication over any appreciable distance is simply not possible. The original antenna-design concepts are generally credited to the physicists Hertz and Marconi. These concepts were based on experiments performed in the late 19th and early 20th centuries. Although considerable progress in wireless communication has been made since that time, the fundamental theories developed by these two men are still valid and still form the basis of modern antenna design.

College-level engineering texts on the subject of antennas and wave propagation contain a considerable amount of complex theory and difficult mathematics. And, if one intends to become a design engineer in this field, such material must, of course, be thoroughly mastered. The average technician, amateur radio hobbyist, CBer, or business radio user, however, does not need to be a design engineer, but he does need to have a good understanding of basic antenna principles.

In this book, all theoretical discussion has been kept as elementary as possible and the use of higher mathematics has been avoided. The text is thus meant to serve primarily as a reference of basic information for anyone desiring to learn the fundamentals of antennas and radio-wave behavior.

Contents

CHAPTER 1

RADIO WAVES AND WAVE PROPAGATION 7

Radio Waves—The Ionosphere—How the Ionosphere Affects Radio Waves—Abnormal Propagation—Polarization—Amplitude—Frequency Bands

CHAPTER 2

ANTENNA FUNDAMENTALS 25

Transmission Line Principles—Basic Antenna Characteristics—Antenna Patterns—Antenna Pattern Development—Frequency and Wavelength—Antenna Length—Directivity and Gain

CHAPTER 3

BASIC ANTENNA DESIGNS 45

Basic Antenna Types—Arrays—Parasitic Elements—Stacked Arrays —Reflectors—Special Antenna Types—Antennas for Microwaves

CHAPTER 4

BASE-STATION ANTENNAS 73

Amateur Antennas—Citizens Band Antennas—Business Radio Antennas—Relay Stations

CHAPTER 5

MOBILE ANTENNAS 99

Mobile Antenna Types—Antenna Loading—Mobile Antennas for Amateur Bands—Mobile Antennas for CB—Mobile Antennas for Business Radio—Marine Antennas

INDEX 123

1

Radio Waves and Wave Propagation

The subject of antennas cannot properly be discussed without first considering the fundamental properties of radio waves—their propagation and behavior as they travel through space. In this chapter, we will examine the basic known facts of wave propagation without employing any lengthy mathematical analysis of the subject. Readers who wish to further pursue a study of this field should realize, however, that a complete mathematical analysis is absolutely essential to a thorough understanding of radio-wave propagation and modern antenna design.

RADIO WAVES

Essentially, radio waves are a form of electromagnetic radiation, the fundamental properties of which are frequency, amplitude, polarization, and direction of travel. Radio waves differ from heat waves, light waves, X-rays, cosmic rays, etc., only in frequency or wavelength. The velocity of these waves depends on the medium through which they travel; in free space it is approximately 186,000 miles per second, or 300,000,000 meters per second. This is also the speed of light in free space, and although the speed is usually thought of as being a constant, there are some velocity changes that occur when a wave passes from one type of transmission medium to another. For example, radio-wave velocity in a medium of copper is less than the velocity in a medium of air. These velocity changes play an important role in radio-wave propagation and will be discussed in greater detail later in the book.

There is an important mathematical relationship between wave velocity, frequency, and wavelength, which is stated by the equation:

$$\lambda = \frac{300{,}000{,}000}{f}$$

where,

λ is the wavelength in meters,
300,000,000 is the speed of light in free space in meters/second,
f is the frequency in hertz (Hz).

Since wavelength is a very important factor in any study of antennas, the reader should thoroughly understand this relationship. Basically the equation states that a low-frequency wave will have a long wavelength, while a high-frequency wave will have a short wavelength. The importance of this fact will be more readily apparent in the discussion of basic antenna design problems in later chapters.

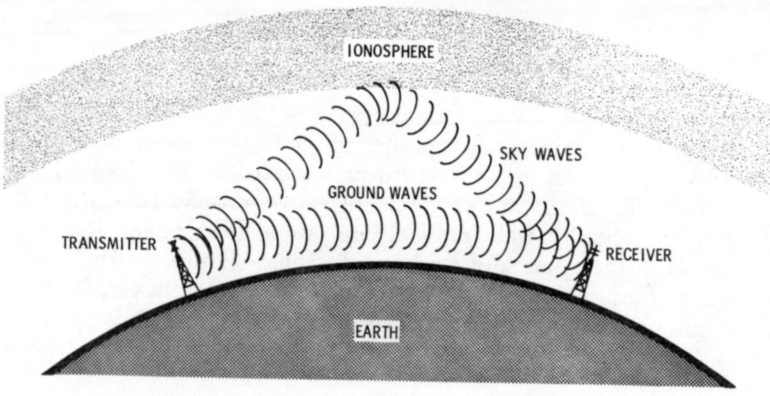

Fig. 1-1. Difference between ground-wave and sky-wave propagation.

In general, radio waves can be classified into two basic types of propagation: (1) ground waves and (2) sky waves. Fig. 1-1 illustrates the primary difference between the two wave types. The ground wave includes all of the components of a radio wave except those affected by the ionosphere or troposphere. This includes what is known as direct, or space, waves (line-of-sight) and ground-reflected waves. Waves reflected back to earth from the ionosphere or troposphere are considered sky waves.

The ground wave (or surface wave) moves along the surface of the earth and actually depends on the earth for a portion of its transmission medium. The sky wave is radiated upward and is affected, more or less according to the frequency, by the ionosphere (ionized regions of the upper atmosphere). Actually, when we speak of ground waves, sky waves, etc., we are referring to portions of the same signal being

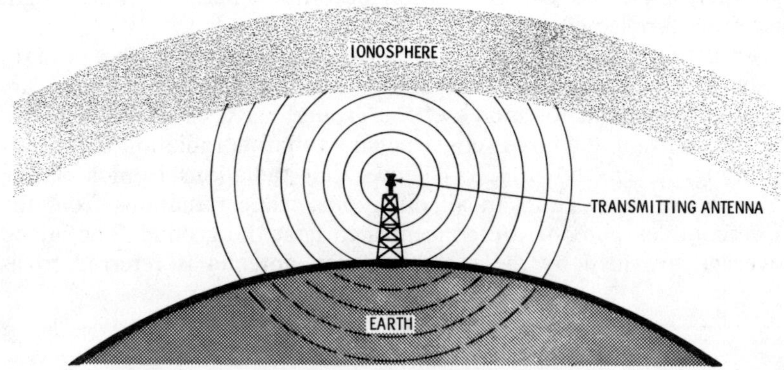

Fig. 1-2. Energy leaves transmitting antenna in the form of wavefronts or shells of energy.

radiated from a particular antenna. A signal leaves a transmitting antenna in the form of spherical shells of energy or wavefronts, as shown in Fig. 1-2. Certain portions of the signal move outward along the ground (the ground wave) and can be detected by a receiving antenna, while other portions move outward and upward into space (the sky wave) and, except for reflection and refraction effects on waves of certain frequencies, are considered wasted energy.

As mentioned earlier, the ground wave depends on the earth's surface for its existence. The earth, however, offers resistance to the flow of currents induced by radio waves and is a limiting factor in the distances that can be covered by primary ground-wave propagation. The ground wave is thus said to be attenuated by the earth's surface. Very sandy soil, for example, becomes less resistive during and immediately after a rain shower but quickly becomes very resistive as it dries. In contrast, the heavy loamy soil of the central United States tends to hold moisture and has low resistance. Thus, ground-wave attenuation is greater in areas having sandy soil than in areas having loamy soil. Ground-wave attenuation is also dependent upon the frequency. At frequencies above 30 megahertz (MHz), ground-wave attenuation is

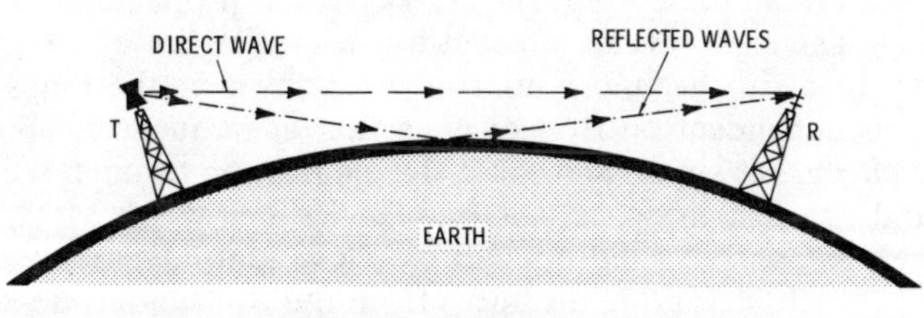

Fig. 1-3. Direct and ground-reflected waves.

extremely high and communication becomes limited to line-of-sight and ground-reflected waves, as illustrated by Fig. 1-3.

Since the ground wave, or certain portions of it, is the primary means of communicating with radio waves, transmitting antennas are designed to extend, or increase, the ground-wave coverage. That is, the transmitting antenna is constructed so that its radiation is primarily along the ground. Fig. 1-4 shows that radiations from a simple antenna are fairly equal in all directions, while radiations from the more complex antenna are concentrated near the ground. The added coverage provided by the more complex antenna is referred to as

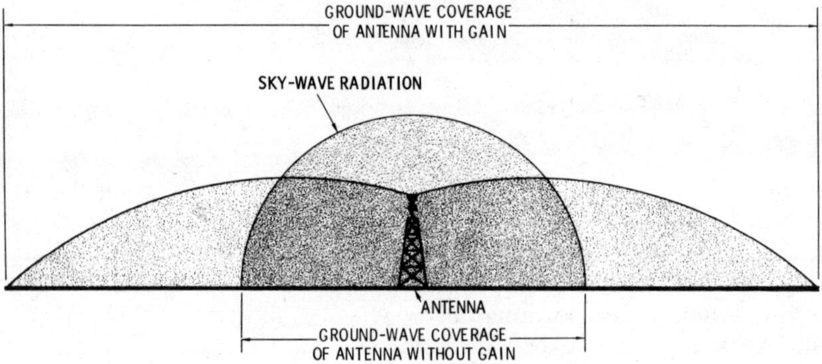

Fig. 1-4. Antenna gain increases ground-wave coverage.

antenna gain, and in later chapters we will be discussing this in greater detail.

THE IONOSPHERE

As mentioned previously, radio waves that travel upward away from the earth are usually considered wasted energy. This, however, depends on the frequency or wavelength of the radiated signal. At frequencies from approximately 3 to 30 MHz, ground-wave attenuation becomes very rapid, limiting coverage with this type of propagation to very short distances from the transmitter. For long-distance communication at these frequencies, sky-wave propagation is used exclusively. The sky wave is affected by several layers of charged or ionized particles in the upper atmosphere known as the ionosphere. This, too, is dependent on frequency so that at frequencies above 30 MHz the effects are small, and these signals pass on through the ionosphere and are lost in space. At frequencies below 30 MHz, radio waves are bent or refracted enough by these ionized particles to be returned to earth at varying distances from the transmitter, depending on the angle of the reflection. At certain frequencies within the 3- to

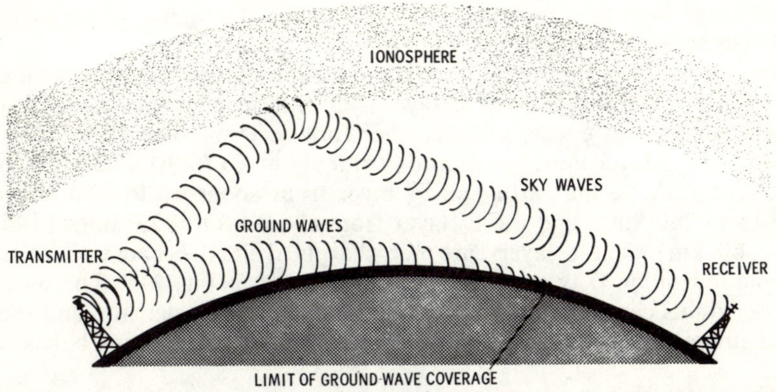

Fig. 1-5. Extended coverage obtained through ionospheric reflection of the sky wave.

30-MHz band, reliable communication at distances of many thousands of miles can be obtained through sky-wave reflection. Fig. 1-5 illustrates the extended coverage that can be obtained through sky-wave reflection.

Ionization of the upper atmosphere occurs primarily because of the very heavy ultraviolet radiation from the sun. Cosmic rays and other radiation from outer space also are a factor. These rays, upon entering the upper atmosphere, encounter molecules and atoms of air and, because of their tremendous velocity and consequent high-energy content, knock electrons from some of the atoms and molecules and create an area of free electrons, positive ions, and negative ions, in a rarefield gas known as the ionosphere.

Scientists have determined that there are at least four layers of ionized particles surrounding the earth. These layers vary in density

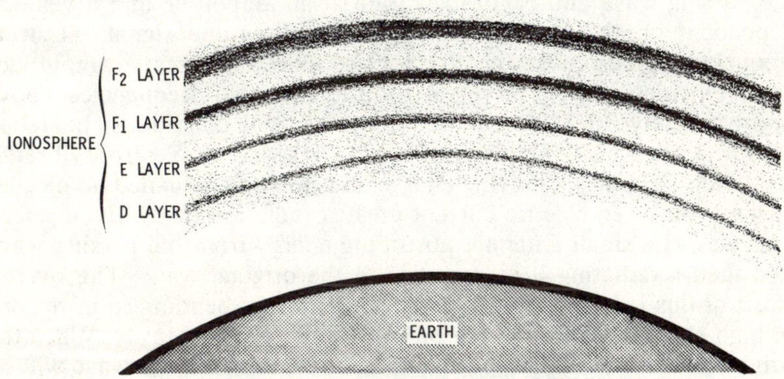

Fig. 1-6. Diagram illustrating the four layers of the ionosphere (usual daytime condition).

and range from approximately 50 to 225 miles (80 to 360 kilometers) above the earth, depending on the time of the year, day, etc. The four layers (Fig. 1-6) during usual daytime conditions are identified as the D, E, F_1, and F_2 layers. The D layer is of little consequence except for some absorption effects and can therefore be ignored in this discussion. The E layer varies in altitude from about 50 to 90 miles (80 to 145 km) above the earth, the F_1 layer from about 90 to 150 miles (145 to 240 km), and the F_2 layer from about 90 to 225 miles (145 to 360 km). These layers are not sharply defined because there is some ionization between them, as can be seen in Fig. 1-6. The term *layer* in this case refers to areas of maximum electron density, and the maximum density of each higher layer is greater than the one beneath it.

At night, during the absence of any direct ultraviolet radiation from the sun, the F_1 layer tends to fade out; the F_2 layer descends and combines with the F_1 layer to form one layer known as the nighttime F layer. The average height of this layer is approximately 150 miles (240 km), which is only slightly affected by seasonal changes. The E layer also tends to fade out at night, and sometimes actually does disappear completely, leaving only the nighttime F layer. Because of this rearrangement of the ionized layers, longer-distance transmission can be obtained at night than during the day.

HOW THE IONOSPHERE AFFECTS RADIO WAVES

As previously stated, the ionized layers of the upper atmosphere act on radio waves of certain frequencies, and this has the effect of changing the direction of travel of the waves. This occurs because of the behavior of the free electrons in the ionosphere when a radio wave enters the region. The free electrons absorb some of the energy from the passing wave and begin to vibrate at an amplitude and a velocity dependent on the wavelength or frequency of the signal. (Lower-frequency signals cause the electrons to vibrate at greater amplitudes and velocities, which accounts for the fact that frequencies above approximately 30 MHz cannot, under normal conditions, be relied on for sky-wave or "skip" transmission.) When an electron vibrates, it actually becomes a moving charge and thus can be called an electric current. Since an electric current creates radiation fields, the electron thus acts as a small antenna, absorbing energy from the passing wave and then reradiating it at an angle to the original wave. The overall effect of this is termed *ionospheric refraction,* or bending from regions of high electron density to regions of lower electron density. When the original wave enters the ionosphere at certain angles, the wave will be returned to the earth at various distances from the transmitter, depending on the angle. Some of the signal will return at distances much

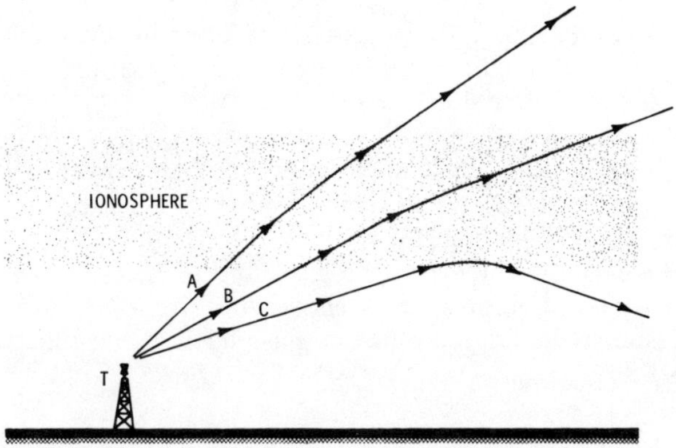

Fig. 1-7. Wave paths in ionosphere for three different angles of incidence.

greater than the maximum possible ground-wave coverage, as shown in Fig. 1-7. The distance between the transmitter and the point where the sky wave returns to earth is called the *skip distance* because, in the area between the limit of ground-wave coverage and the point where the sky wave returns, the signal cannot be received.

To better understand how ionospheric refraction occurs, let us briefly examine the laws of refraction as they apply to the field of optics and then correlate this information with the behavior of radio waves. Fig. 1-8 shows an ordinary sheet of glass. A light ray traveling through the air enters the glass at point A, passes through the glass from point A to point B, and returns to the air.

In order to explain this action, a vertical line known as a *normal* (N_1) is drawn at point A and a second vertical line or normal (N_2) is drawn at point B. Note that when the light ray hits the glass, it makes an angle (angle 1) with the normal (N_1). However, because

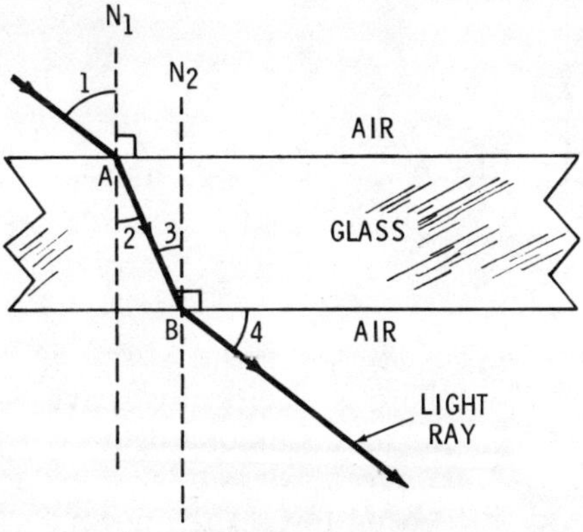

Fig. 1-8. Path of light ray through glass.

light travels more slowly in glass than it does in air (glass is more dense than air), the light ray is bent toward the normal so that angle 2 is smaller than angle 1.

This is a basic characteristic of refraction; a light ray going from a less dense to a more dense medium will be bent toward the normal. Conversely, a light ray going from a more dense medium to a less dense medium will be bent away from the normal. This is indicated at point B where the ray leaves the glass and reenters the air.

A somewhat different case is shown in Fig. 1-9. Here the light wave is assumed to originate in the glass itself. Note that light wave

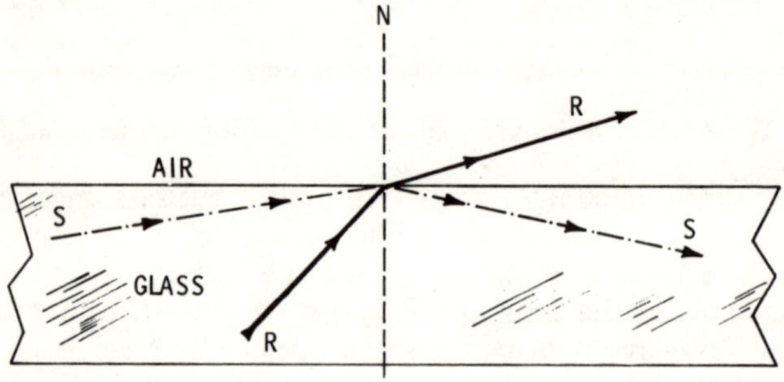

Fig. 1-9. Paths of two light rays originating inside glass.

R strikes the surface of the glass and passes on to the air as before. Light ray S, however, does not escape to the air but is refracted in such a manner that it is retained inside the glass. The point to be learned here is that at a certain angle, a light ray originating inside the glass appears to be reflected from the boundary of the two mediums and to stay inside the glass.

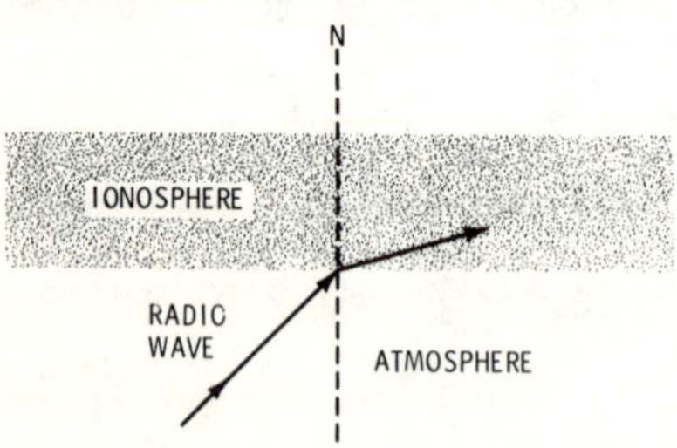

Fig. 1-10. Ionosphere causes radio wave to bend away from normal.

Because the ionosphere is electrically active, its refraction of a radio wave differs from the refraction of a light ray by a sheet of glass. As shown in Fig. 1-8, the greater-density glass reduces the light ray's velocity and causes it to bend toward the normal (N_1). In contrast, the electrically dense ionosphere bends the radio wave away from the normal (N) in Fig. 1-10. Fig. 1-11 shows this in greater detail as a radio wave front (X-Y) enters the ionosphere layer. Point X on this wave front enters the ionosphere layer first and is accelerated by its reaction with the free electrons. The wave front is thus turned or bent away from the normal (N) as indicated by the steps of Figs. 1-11A, B, and C. In Fig. 1-11C, point X has reached the point of greatest electron density and the reaction and bending are also at a maximum. When the electron density produces enough bending, the radio wave turns back toward the earth, as illustrated by Fig. 1-11D. Actually, to achieve complete refraction back toward the earth, as in Fig. 1-7 (wave C) and Fig. 1-11D, may require more than one ionosphere layer. While the electron densities are greater during the daytime, the greater height of the ionosphere layers makes the complete

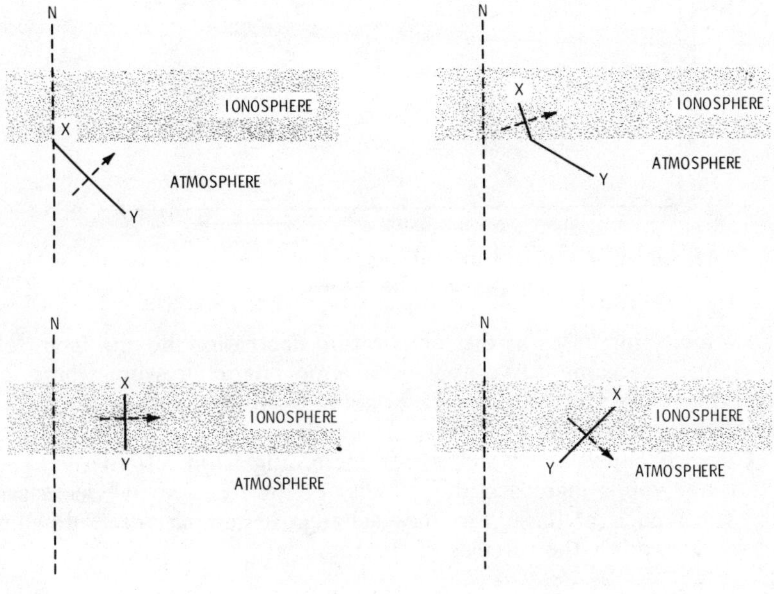

Fig. 1-11. Radio wavefront (X-Y) entering and leaving the ionosphere.

refraction condition less likely. Therefore, radio-wave reflection, produced by ionospheric refraction, is more apt to occur at night. For this reason, distant communication using radio waves is commonly carried out during nighttime hours. Amateur radio operators using frequencies below 30 MHz are particularly fond of working their "rigs" at night.

Dependence upon ionospheric reflection has been reduced in recent years by man-made communications satellites. Such communications satellites are permanently located in space to provide dependable radio-wave reflections at frequencies far above the 30-MHz limit of the ionosphere.

ABNORMAL PROPAGATION

Several abnormal forms of wave propagations occur within the atmosphere itself. In traveling from a more dense into a less dense medium, a radio wave or light wave is refracted so that it bends away from the normal, but this is only part of the picture. The paths of many radio signals are curved, since they move through the atmosphere with different velocities. Their actual velocity depends on a number of factors, including temperature, atmospheric pressure, and the amount of water vapor in the air. Curve A in Fig. 1-12 shows that

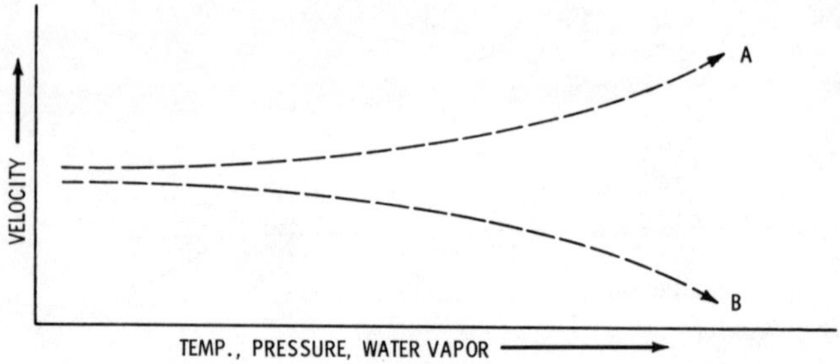

Fig. 1-12. Effect of temperature, pressure, and water vapor on the velocity of radio waves.

the velocity increases as the temperature decreases, moving from left to right. This same curve applies to atmospheric pressure where, as the atmospheric pressure decreases going from left to right, the velocity of the signal again increases.

Curve B shows the water vapor increasing from left to right. As the water vapor increases, the velocity of the radio signal decreases. Thus, for each of these variables (change in temperature, pressure, or water vapor), the velocity of the radio waves changes.

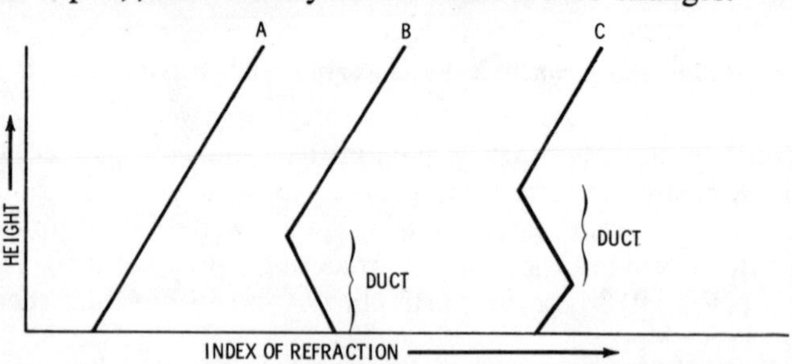

Fig. 1-13. Index of refraction versus altitude curves.

Considering all of these factors, it is clear that the velocity of radio waves will also vary with altitude, since temperature, pressure, and water vapor all change with altitude. Under normal atmospheric conditions, this variation or change is quite small. However, there is an increase in the velocity of radio waves as the altitude increases. This causes a small curvature in the downward direction of a radio wave traveling over the surface of the earth. Instead of the line-of-sight propagation distance of 25 miles (40 km) originally expected with frequencies above 30 MHz, normal atmospheric refraction extends that distance to better than 40 miles (65 km), as indicated by Fig. 1-14.

The temperature and water vapor in the atmosphere undergo constant change; usually this change is such that the index of refraction will vary with the altitude, as shown in Fig. 1-13A. This is the standard, or normal, change in the index of refraction as a function of the height or altitude above sea level.

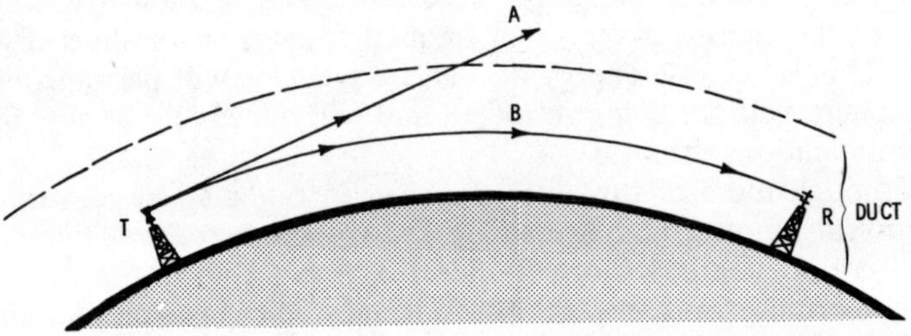

Fig. 1-14. Increase in transmission range due to duct effect.

There are conditions, however, where the change in the refraction index is not normal, bringing about abnormal or anomalous propagation. A duct or a region of special characteristics is formed under these conditions, and the duct will carry the radio waves in a manner similar to a waveguide.

Normally, as the altitude is increased, the temperature will decrease. Also, as the altitude increases, the amount of water vapor decreases. There are special circumstances, however, where the temperature first decreases as the altitude is increased and then suddenly the temperature will increase again. This is an abnormal condition known as temperature inversion. It is also possible, under particular circumstances, for the water vapor content of the atmosphere to increase at a certain point above the surface of the earth.

When these things happen, the unusual conditions form a duct; Fig. 1-13B shows that the index of refraction changes in an unusual way as the altitude is increased, and then the change becomes normal for a further increase in altitude. In some cases, as in Fig. 1-13C, the duct occurs between two specific levels of altitude. In any case, the formation of the duct is unusual, and it affects radio waves in an unusual manner.

Fig. 1-14 shows radio waves starting out at transmitter T directed toward receiver R. Under normal conditions, radio waves would travel in about a straight line, as shown at A. However, because of the duct effect, the radio wave is bent in its travel in path B. Hence, energy from the transmitter reaches receivers at a greater distance in this abnormal fashion. This duct transmission effect may increase the transmission range of certain types of radiocommunciations by many times.*

Other types of transmission effects that are not abnormal but are unusual include reflections from things other than the ionosphere. For example, passive satellites have been used for communications in which energy is sent from the transmitter, bounced off the satellite, and returned to the earth. Tropospheric scatter propagation is another means of extending the range of certain types of radio transmission. Most of the energy at very high frequencies is sent for short distances only. This is because energy at high frequencies will pass through the ionosphere without being refracted and will effectively be lost for any communications use.

Although most of this energy is lost in outer space, a very small but useful amount is deflected by the troposphere, and this scatters the energy beyond the horizon. Thus, it is possible to use frequencies for communications beyond the normal radio horizon that usually would not be used for long-range transmission. Tropospheric scatter propagation is not the same as reflection from the ionosphere because tropospheric scatter depends on certain atmospheric conditions that do not exist all the time. Reflections from the ionosphere, however, are possible on a continuing basis.

POLARIZATION

A radio wave consists of an electric-field component and a magnetic-field component. These are at right angles, as shown in Fig. 1-15, and they both vary as a function of time. The direction of wave travel here is shown by the direction of the arrowhead. The magnetic field can be considered as the horizontally moving sine wave representing variations in the magnetic field, while the electric field can be considered as the vertically moving sine wave representing variations in the electric field. All radio waves have these two components and can be diagrammatically shown in a variety of ways.

---

* Normal radio waves do bend in their travel and will be received somewhat beyond normal line-of-sight in many cases.

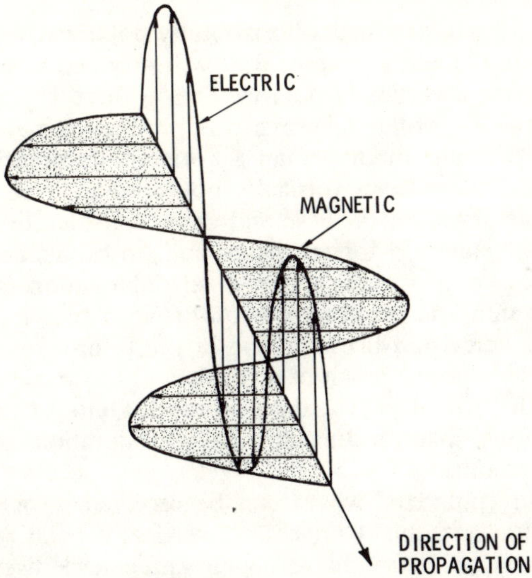

Fig. 1-15. Diagram illustrating electric- and magnetic-field components of a radio wave.

In Fig. 1-16, for example, consider that the wave is traveling out of the paper toward your eyes. The magnetic-field vector is now shown vertically, and the electric-field vector is shown horizontally. The direction of the electric field in a wave determines the polarization of that wave. Where the electric vector is horizontal, as in Fig. 1-17A, the radio wave is known as a *horizontally polarized* wave. Where the electric vector is vertical, as in Fig. 1-17B, the radio wave is known as a *vertically polarized* wave. The electric field and the magnetic field are always at right angles to each other.

Polarization of radio waves depends on the position of the transmitting antenna relative to the earth. When the transmitting antenna

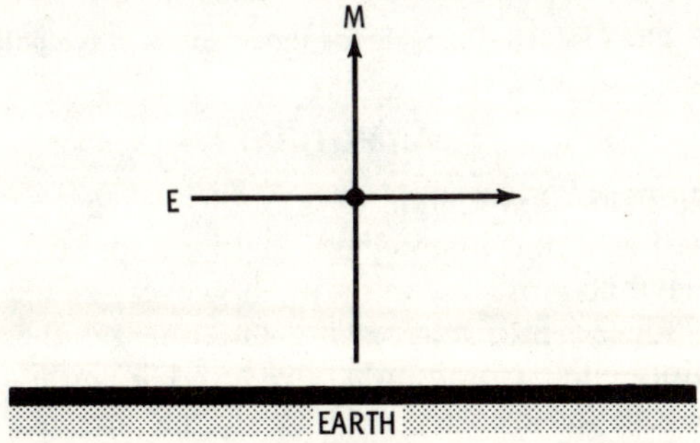

Fig. 1-16. Vector representation of wavefront.

is horizontal, it will produce a horizontally polarized wave; and when the transmitting antenna is vertical, it will produce a vertically polarized wave. With a vertically polarized wave there is less attenuation; hence, a vertically polarized wave will produce somewhat more energy at the receiving antenna than a horizontally polarized wave. At the same time, however, a vertically polarized wave is affected more by noise. Thus, because of these and other considerations, horizontal polarization is standard for television and fm broadcasting, although a combination of horizontal and vertical polarization (known as *circular* polarization) is permitted for fm broadcasting and may also be permitted for television broadcasting in the future. In other types of services, noise is not as much of a factor and the vertical transmitting antenna, with vertical polarization, is used. Quite often with mobile communications, space and mechanical considerations require the use of vertical antennas.

Horizontally polarized waves can be received on a vertically oriented receiving antenna. Conversely, vertically polarized waves can be received on a horizontal receiving antenna. However, such arrangements are not advisable, since the electric field attenuates rather fast and the magnetic field of a vertically polarized wave will induce very little energy into a horizontal receiving antenna.

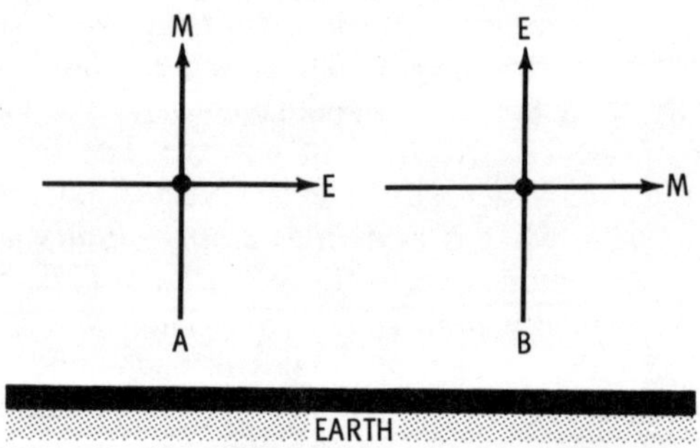

(A) Horizontally polarized wave. *(B) Vertically polarized wave.*

Fig. 1-17. The electric-field vector determines wave polarization.

AMPLITUDE

The electric-field component of a radio wave has an amplitude measured in voltage, and its magnetic-field component is measured in oersteds. Somewhat more meaningful, however, is the electromotive force (emf) induced into a receiving antenna by these electric and magnetic components. Commonly measured in millivolts per meter (mV/m), the emf induced by the electric and magnetic components is referred to as the *field strength*. Although relative to the wave amplitude, the field strength is not an actual measure of the wave or its component amplitudes. Broadcast stations identify their primary service areas by the field strength that provides adequate reception. For example, a field strength of less than 50 mV/m will not provide adequate television reception. Naturally there are areas with television reception of less than 50 mV/m, but the resulting picture usually has poor resolution, snow caused by noise, and little or no color. Since advertisers do not accept such television as being adequate, the broadcaster must base his advertising upon that area having an adequate field strength. In contrast, the standard a-m broadcast station provides adequate service in low-noise areas with field strengths of 10 mV/m or less. Other communication services find even lower field strengths sufficient.

FREQUENCY BANDS

To summarize and to illustrate the importance of frequency in relation to radio-wave propagation, the following paragraphs list the propagation characteristics of the various frequency bands from very low to superhigh frequencies.

1. *Very low frequency* (vlf) 10 to 30 kHz—Radio waves at these frequencies are very reliable for long-range communications. Distances of many thousands of kilometers are possible in the daytime. Attenuation of the ground wave is very small, and the sky-wave reflection is good. However, the antenna systems required for transmission in this frequency range must of necessity be quite large and expensive.
2. *Low frequency* (lf) 30 to 300 kHz—The ground-wave coverage in this band is somewhat less than in the vlf band because of the increased attenuation by the earth's surface. Since the ionosphere, in the daytime, does not reflect these waves as readily as those of the vlf band, sky absorption begins to be a factor, especially at the upper end of this band. Instead of the thousands of kilometers of vlf daytime coverage, frequencies of this band are reduced to hundreds of kilometers of coverage. At night, it is possible to obtain longer-range transmission with ionospheric reflection of the sky wave.
3. *Medium frequency* (mf) 300 to 3000 kHz—Some aircraft and weather communications use the lower end of this band because of its dependability and distance range. Somewhat better known is the 550- to 1600-kHz portion used for standard a-m broadcasting. Reliable ground-wave coverage is possible for distances up to 100 miles (160 km) from the transmitting antenna. And while even greater coverage is usually possible at night with sky-wave reflections, the broadcaster is basically interested in the

30- to 40-mile (50- to 65-km) radius defining the primary service area.

4. *High frequency* (hf) 3 to 30 MHz—This band, like many of the other frequency bands, has been reevaluated in recent years. Originally used for distance communications with sky-wave reflection, this band is now used for a number of localized services. Of primary interest among these services is the Class D Citizens band (CB) having 22 channels between 26.965 MHz and 27.225 MHz. Another channel, at 27.255 MHz, is shared with radio-control devices such as model airplanes. Amateur radio also uses portions of this band, depending upon sky-wave propagation and ionospheric reflection for long-distance communications. With proper base and mobile unit antennas, 25-mile (40-km) coverage is possible with a transmitter power of 15 watts.
5. *Very high frequency* (vhf) 30 to 300 MHz—This band includes the commercial fm and vhf television channels. Since space-wave propagation (line-of-sight) is the principal means of communication, transmitting antennas for this band are usually mounted high above ground level. While there may be some atmospheric refraction, ionospheric reflection seldom occurs in this band. Commercial fm stations use frequencies between 88 and 108 MHz, while television stations use channels from 54 to 72 MHz, from 76 to 88 MHz, and from 174 to 216 MHz. The various aviation services such as air-traffic control and radio-navigation use frequencies between 108 and 136 MHz. Marine radio also uses frequencies of this band as do public-safety services, such as police and fire departments, and many types of industrial or business radio services.
6. *Ultrahigh frequency* (uhf) 300 to 3000 MHz—Propagation in this band is possible somewhat beyond the optical line-of-sight due to the increasing refraction effects of the atmosphere. The actual distance that can be covered depends on the transmitting and receiving antenna heights. There is also an increase in the atmospheric attenuation due to moisture, etc. The most noteworthy usage within this band is uhf television, with channels between 470 and 806 MHz. The upper limits of this band, above 1000 MHz, are referred to as microwaves and are used in radar, microwave cooking, relaying of telephone conversations, television signals, telemetry data, etc.
7. *Superhigh frequency* (shf) 3 to 30 gigahertz (GHz) (3000 to 30,000 MHz)—At these frequencies, especially those above 10 GHz, atmospheric absorption due to moisture and precipitation becomes very considerable. Presently, this band represents the upper limit of frequencies having practical use in radio-wave

communications. Before higher frequencies can be generated and transmitted, new techniques and devices will be needed. Some experimental work with *maser* and *laser* beams is now leading in that direction. In the future, it may be common to communicate on frequencies as high as light waves (about 5000 GHz).

2

Antenna Fundamentals

Antennas are a key part of all forms of radio-wave transmission and reception. The simple dipole, for example, is the proverbial "workhorse" and is used in all forms of radio-wave communications. Half-wave dipoles and folded half-wave dipoles are fundamental elements in most antenna systems. For special applications, other types such as the fan dipole and the Yagi are important for increased gain.

Electrically, the dipole antenna is equivalent to an open-circuited transmission line. Fed by a generator, the dipole appears to that generator as a resonant circuit. Since the wires are in line, rather than parallel as with the usual transmission line, the current produces radiating electric and magnetic fields. These radiating fields are responsible for the radio waves moving out away from the transmitting antenna. In reverse, these radiating fields induce current into the receiving dipole antenna.

The sharpness of the response of the dipole antenna is similar to that of an ordinary low-frequency lumped inductance and capacitance type of resonant circuit, depending on the type of construction used in the dipole. For example, if the dipole is made of thin copper wire, it behaves like a very sharply tuned circuit with a high Q. Such an antenna can operate only over a narrow band of frequencies. When the dipole conductors are increased in diameter, the frequency response will be broadened.

TRANSMISSION LINE PRINCIPLES

An antenna can be considered a special type of transmission line designed to radiate or receive energy. In either case, the characteristics of a transmission line can be compared to basic antenna behavior.

Essentially, a transmission line is a pair of wires designed to transfer energy from one point to another. For example, a transmission line is used between the radio transmitter and the antenna mounted on a tower next to the transmitter. The purpose of the transmission line is to take energy from the transmitter, carry it up the tower, and feed it to the transmitting antenna. In the same manner, a television receiver transmission line connects the antenna on the roof to the receiver in the house. It carries the energy picked up by the antenna down to the television receiver.

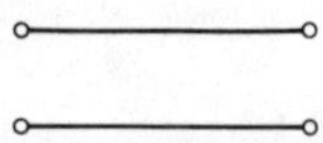

(A) Transmission line.

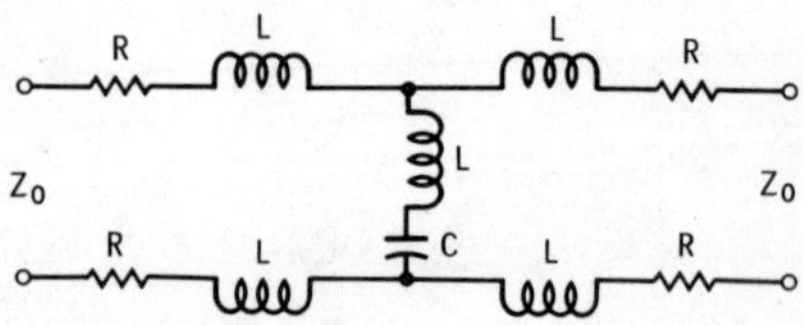

(B) Electrical equivalent to a transmission line in distributed elements.

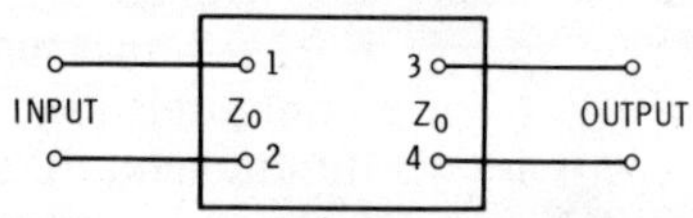

(C) Four-terminal equivalent to a transmission line.

Fig. 2-1. Transmission-line characteristics.

While the fundamental purpose of the transmission line is to carry electrical energy, it has characteristics that are important to the understanding of antennas. Mechanically, the transmission line consists of two parallel wires as shown in Fig. 2-1A. Electrically, this becomes a complexity of distributed inductances, capacitances, and resistances as shown in Fig. 2-1B. But this is too complex for our study and it is sufficient to say that the transmission line has a *characteristic impedance,* Z_0. Since this is the result of the various distributed elements shown in Fig. 2-1B, a transmission line that is 100 feet (30.48 meters) in length has the same characteristic impedance as a 1.0-ft (30.48-cm) line. Thus, the transmission line can be looked upon as a four-terminal network, as shown in Fig. 2-1C. Both ends of the transmission line are identical and the input impedance, Z_{in}, equals the output impedance, Z_{out}. In similar manner, the input impedance across terminals 1 and 2 of the four-terminal network equals the output impedance across terminals 3 and 4.

Basic electronic theory states that a generator transfers a maximum amount of energy when the load impedance, Z_L, equals the internal impedance, R_g, of the generator. For example, when generator G in Fig. 2-2A has an internal impedance of 75 ohms and develops a voltage of 150 volts, a current of 1.0 ampere is produced through a

Z_L of 75 ohms. Under these conditions, the voltage across the load impedance will be 75 volts, and 75 watts of energy will be delivered to the load. Reducing the load impedance to 70 ohms allows the current to increase to 1.031 amperes, delivering 74.4 watts to the load. Conversely, an increase in the load impedance will decrease the amount of current and energy transferred. It is as though the decreased or increased load impedance rejects a portion of the energy and sends it back to the generator.

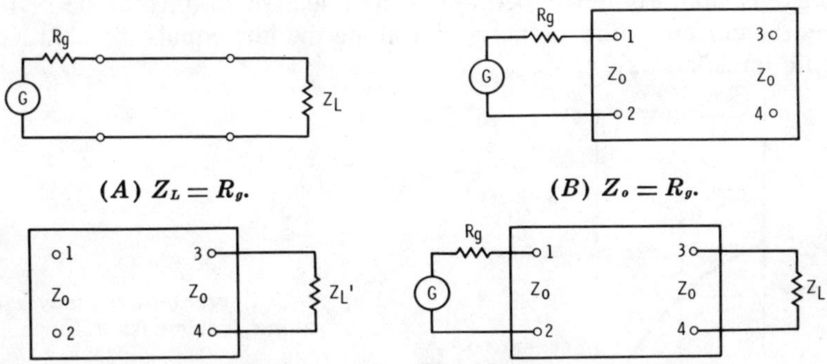

(A) $Z_L = R_g$. (B) $Z_o = R_g$.

(C) $Z_L' = Z_o$. (D) $Z_L' = Z_o = R_g$.

Fig. 2-2. Conditions for maximum transfer of energy between generator and load.

In Fig. 2-2B, the load impedance, Z_L, has been replaced by the characteristic impedance, Z_o, of the four-terminal network. A maximum amount of energy is transferred when Z_o equals the internal impedance of the generator. The four-terminal network becomes the generator when a load impedance is placed across output terminals 3 and 4 in Fig. 2-2C. And here again, the load impedance, Z_L', must equal the generator impedance (in this case, Z_o) for the maximum transfer of energy to occur. A load impedance not equal to Z_o rejects a portion of the energy delivered and sends it back through the four-terminal network to the generator in the complete circuit of Fig. 2-2D. This complete circuit is equivalent to a radio transmitter feeding energy through a transmission line to an antenna.

Current moving along a conductor develops a magnetic field proportional to the current amplitude. That magnetic field moves along the conductor and tends to oppose any change in current amplitude. So, as an alternating current flows into the input of a transmission line (a conductor), the current at points along the line is the result of voltages applied at some earlier instant. For example, in the drawing of Fig. 2-3 the input current is shown at A as having an amplitude of 1.0. One-eighth of the time of one cycle before this instant, the current had an amplitude of 0.707 as shown at point B. The distance between points A and B is the distance a wave will travel during one-eighth cycle, or one-eighth of a wavelength (λ/8). Similarly, point C on the line has a current value equal to that at the line input one-fourth of a wavelength (λ/4) before the instant at point A. In this manner, the various current values move along the transmission line as indicated. One point that must be understood is that the wave shown in Fig. 2-3 is moving from left to right. The point B value was at point A just an instant before, while the value at point C was at point B. A voltage wave is similarly developed along the line. The voltage value divided by the current value at each point along the line equals the characteristic impedance, Z_o.

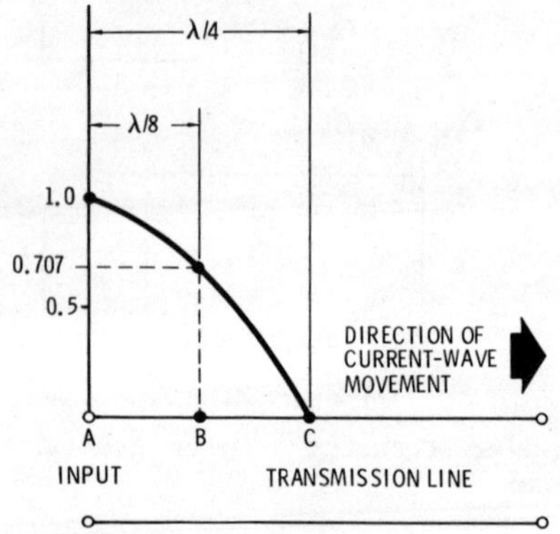

Fig. 2-3. Current-wave movement along transmission line.

Thus, the transmission line tends to be a constant-voltage/constant-current device. When a 75-volt wave is applied to a line having a characteristic impedance of 75 ohms, its current wave has a maximum value of 1.0 ampere. Terminating such a line with a load impedance of 75 ohms will absorb all of the energy delivered. A larger load impedance of 150 ohms will absorb a very small portion of the energy delivered, and much of the current and voltage waves will be reflected

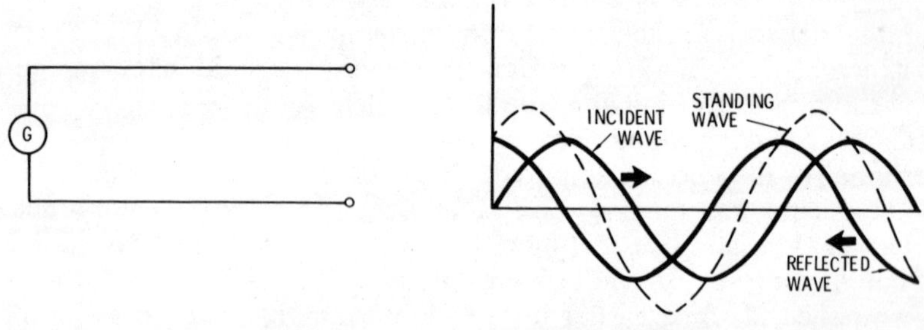

(A) Open-ended transmission line.

(B) Standing wave is effective sum of incident and reflected waves.

Fig. 2-4. Development of standing wave.

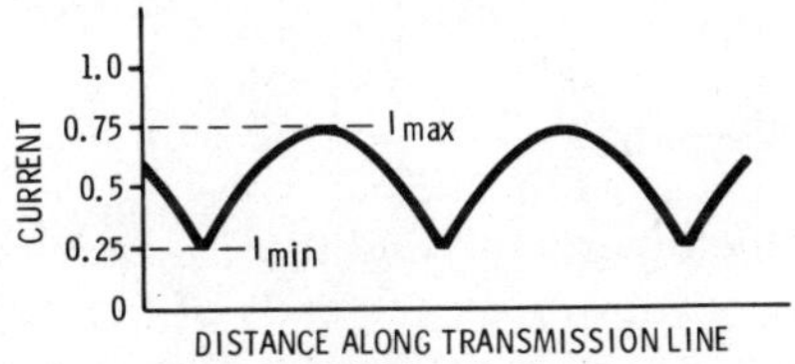

Fig. 2-5. Illustration of standing-wave ratio. In this example, swr = I_{max}/I_{min} = 0.75/0.25 = 3 to 1.

back into the transmission line. This is shown in Fig. 2-4A, where the load impedance is made infinite by leaving the receiving end of the transmission line open. Such an open end will absorb none of the delivered energy. At this open end, the current will drop to a minimum value. However, the change in current induces a voltage opposing that change. As a result, the incident (or initial) wave of energy is opposed by an energy wave moving in the opposite direction. Thus, the incident wave moves from left to right and the reflected wave moves from right to left as shown in Fig. 2-4B. A meter placed within the line at any point will indicate the effective sum of the incident wave and the reflected wave. That effective sum also takes the form of a wave. However, the effective sum wave does not move (although it fluctuates in intensity) and is referred to as the *standing wave* (Fig. 2-4B). Although the standing wave does have both positive and negative maximums, referred to as *nodes,* a meter indicating effective values will only indicate points of maximum and minimum as shown in Fig. 2-5. Since the load impedance (even infinite or zero loads) always absorbs some of the energy delivered, the reflected wave is never exactly equal and opposite to the incident wave, and the standing wave is never zero. The relationship of the standing wave maximum to its minimum is known as the *standing-wave ratio* (swr). The swr is also equal to the ratio of the load impedance to the line characteristic impedance:

$$\text{swr} = Z_L/Z_o \text{ or } Z_o/Z_L$$

The standing wave, with its amplitude variations, produces magnetic and electric radiations–radio waves. Such radiations from a transmission line represent lost energy. For this reason, the standing wave is an undesirable factor in transmission lines. Radiations developed by the usual parallel-wire transmission line tend to oppose each other and are not propagated to any appreciable distance. Even less propagation occurs with a coaxial transmission line having one conductor concentrically placed within a tubular conductor.

BASIC ANTENNA CHARACTERISTICS

Fig. 2-6 demonstrates how the transmitter, or radio-frequency generator, uses a transmission line as an antenna. In Fig. 2-6A, the transmitter is connected to a short section of transmission line. Since the line is open-ended, no energy is absorbed and a standing wave is developed. The transmission line wires are close together and parallel to each other. Each wire creates a field that is cancelled by that of the other wire, and there is little radiation of energy.

If, however, the ends of the transmission line are partially opened, as in Fig. 2-6B, there will be some radiation of energy in the direction indicated by the arrow. Electric and magnetic fields radiate perpendicularly from the wires, so that with nonparallel wires the cancellation of fields is incomplete.

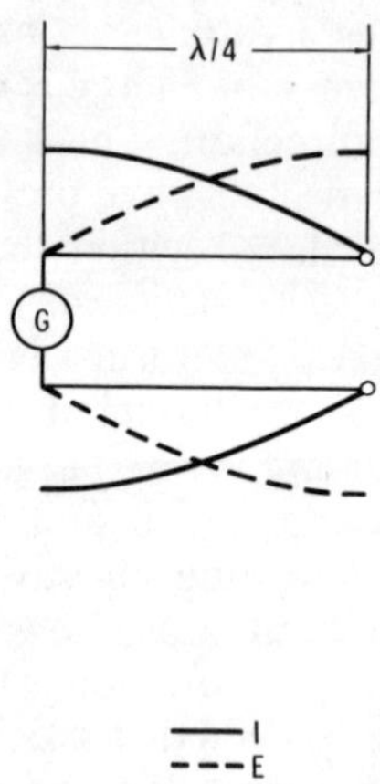

(A) Generator connected to quarter-wavelength transmission line.

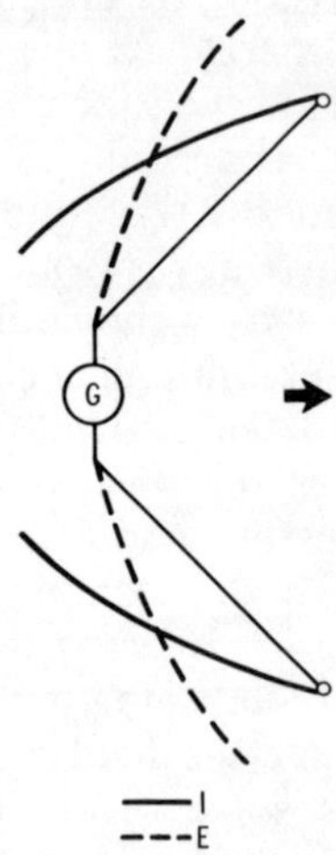

(B) Ends of transmission line partly opened.

(C) Transmission line completely opened forming half-wavelength antenna.

Fig. 2-6. Evolution of antenna from transmission line.

In Fig. 2-6C, the transmission line is opened all the way so that the two wires are now in a straight line. Now there will be maximum radiation of energy, since the fields created about the two wires do not oppose each other. This represents an actual antenna.

As shown in Fig. 2-7, the two sections of transmission line (each one-quarter wavelength long) form a half-wave antenna. This antenna has a voltage maximum and a current minimum at either end. This represents the fundamental half-wave dipole antenna. Having low-voltage and high-current standing-wave values at its input, the dipole antenna represents a series-resonant circuit with a low impedance. Actually, the input impedance at resonance will be close to the characteristic impedance of the transmission line—72 ohms.

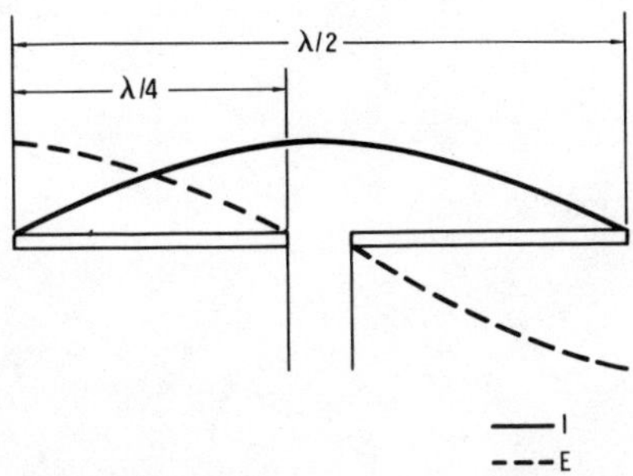

Fig. 2-7. Standing current and voltage waves on half-wavelength antenna.

ANTENNA PATTERNS

The antenna pattern is a graphic means of defining the directional characteristics of an antenna. Also referred to as a *polar* diagram, the antenna pattern plots those points having equal wave intensities at various angles about the antenna. The antenna pattern can also be a plot of relative wave intensities.

Initially, let us assume that the antenna is located at the center of a sphere. Every point on the inner surface of that sphere would be acted upon by equal wave intensity if the energy were radiated equally in all directions. Radiation from the dipole is perpendicular to the conductor and is equal in all directions on a plane through the center of the dipole (Fig. 2-8A). On a plane parallel to the dipole conductors, the radiation pattern is that of a figure eight, as seen in Fig. 2-8B. Together, the patterns of Figs. 2-8A and 2-8B have the three-dimensional form of a doughnut as shown in Fig. 2-8C. However, the interest is usually in the vertical or horizontal patterns. The horizontal pattern provides knowledge of direction and area covered by the radiation or reception of the antenna. Vertical patterns give an indication of the amount of radiation being lost as sky waves. The same antenna patterns apply for an antenna used for either transmitting or receiving.

The directivity of an antenna varies with its electrical length; Fig. 2-9 shows the characteristic radiation patterns of a dipole antenna for four different frequencies of operation. The antenna is horizontal in this case, with maximum radiation or reception being in the directions shown. The patterns are doughnut-shaped, as in Fig. 2-8C, so they can be thought of as extending into and out of the pages of this book.

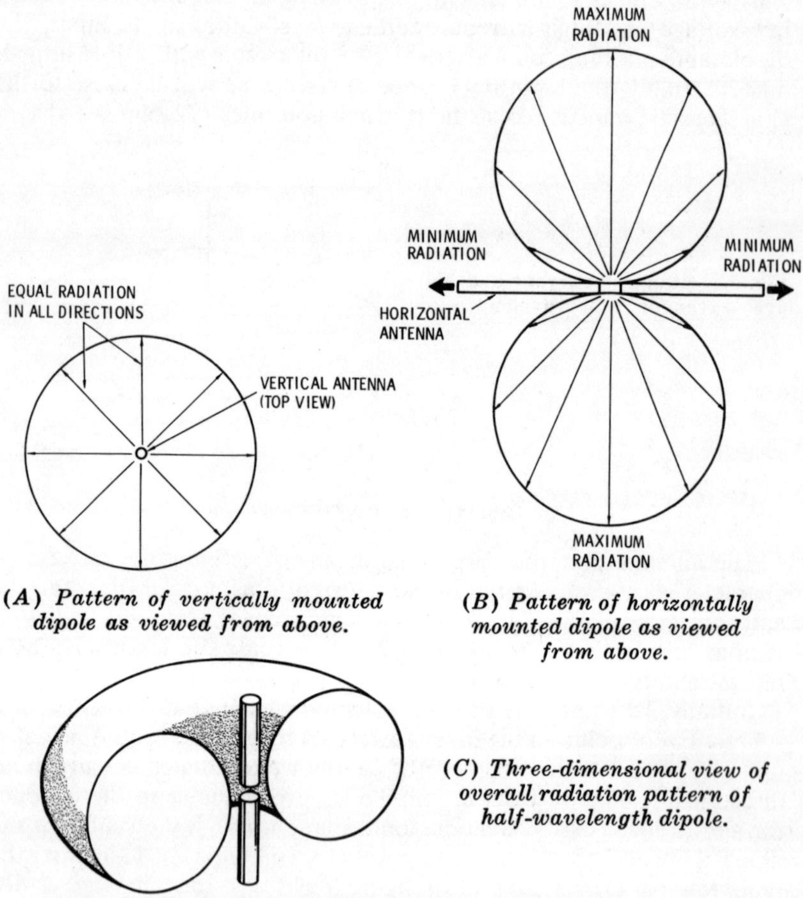

(*A*) *Pattern of vertically mounted dipole as viewed from above.*

(*B*) *Pattern of horizontally mounted dipole as viewed from above.*

(*C*) *Three-dimensional view of overall radiation pattern of half-wavelength dipole.*

Fig. 2-8. Radiation patterns of half-wavelength dipole antenna.

Fig. 2-9A shows the directional characteristics of the typical half-wave antenna previously shown in Fig. 2-8B. When the operating frequency is doubled, the antenna effectively becomes a full wavelength, and the pattern changes (Fig. 2-9B). Note that there are now four loops, two in the forward direction and two in the reverse direc-

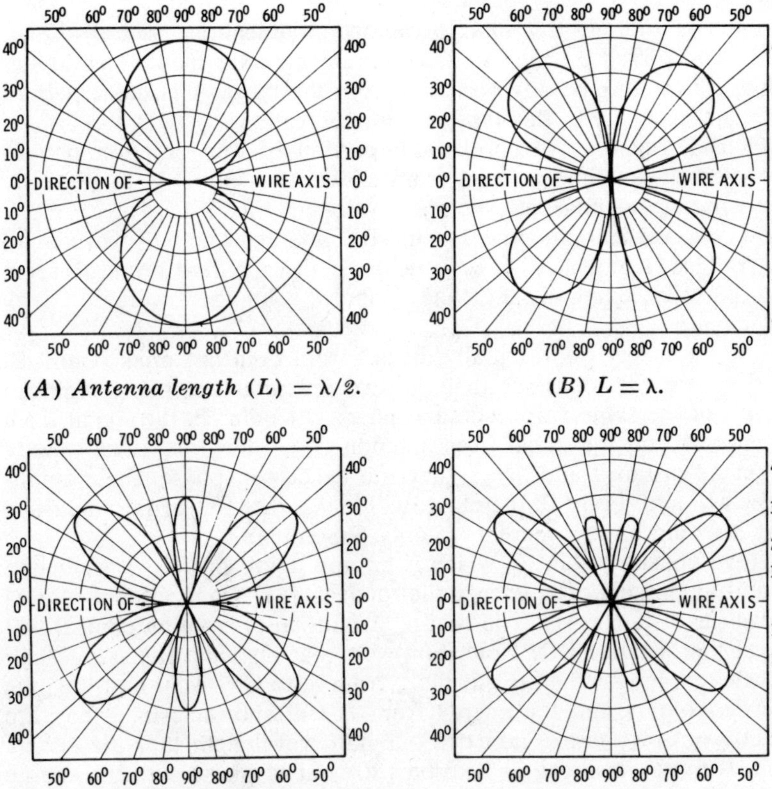

(*A*) *Antenna length* (*L*) $= \lambda/2$.

(*B*) $L = \lambda$.

(*C*) $L = 1\frac{1}{2}\lambda$.

(*D*) $L = 2\lambda$.

Fig. 2-9. Typical radiation patterns for the same dipole antenna at four different frequencies of operation.

tion. There is maximum signal transmission or reception in each of the four directions instead of in two directions as before.

Fig. 2-9C shows the same antenna operated at three times the original frequency. Now there are three forward loops and three rear loops, and the antenna is effectively one and one-half wavelengths long. In Fig. 2-9D, the antenna is operated at four times the original frequency and is now two full wavelengths long. The antenna now has four forward loops and four rear loops. In general, the higher the frequency of operation is for a given antenna length, the more loops there are developed.

ANTENNA PATTERN DEVELOPMENT

A vertical antenna will have a circular pattern when viewed from above the antenna, as shown in Fig. 2-8A. A typical use of this type antenna is in two-way mobile radiocommunications where a base station in a fixed position must communicate with many mobile stations, which may be any direction from the base. A circular pattern is thus a necessity in this type of communication.

Considering only the horizontal portion of the radiation pattern of a vertical antenna as seen from above, it is possible to illustrate how special antenna patterns are developed. Fig. 2-10, for example, shows two vertical antennas (A and B) mounted one-half wavelength apart. This is indicated as $\lambda/2$, which is the standard notation for a one-half wavelength distance, the symbol λ being the lowercase Greek letter *lambda*.

Point P is any point equally distant from both antennas A and B. In this example, consider that the two antennas are being fed with signals of the same amplitude and phase. At point P, the signal from antenna A and the signal from antenna B will have to travel exactly the same distance and, since they are exactly in phase, their amplitudes will add. Point P, therefore, will have twice the signal amplitude that it would have if there were only one antenna.

Now, consider point Q which is on any point in line with antenna A and antenna B. The energy that arrives at Q from antenna A will travel the distance AQ. The energy arriving at Q from antenna B will travel the distance BQ. Note, however, that distance BQ is one-half wavelength longer than the distance from A to Q. Because of this, the energy from A and the energy from B will subtract since they are exactly out of phase. The extra one-half wavelength that the energy from B must travel causes it to be 180° out of phase with the energy from antenna A. Hence, the two signals must subtract, and there will be no resultant energy at point Q.

This is a simple example that illustrates basically how antenna radiation patterns are developed. While this example utilized the pattern

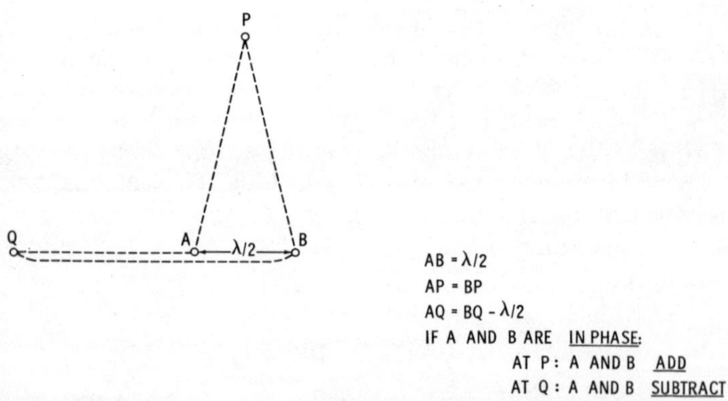

Fig. 2-10. Drawing illustrating how special antenna patterns are developed.

of two vertical antennas, the same considerations apply to the radiation patterns of all antennas. The single antenna is considered to be several smaller radiating elements and a summation of their effect at different angles is made. Because such calculations require careful attention to standing-wave amplitude and polarity, angles, trigonometric functions, etc., the actual calculations are beyond the scope of this book.

Another method of determining the radiation pattern of an antenna involves actual laboratory or field measurements. In this method, the antenna is rotated to different angles, and relative field-strength readings are made at some distant point. With fixed broadcast antennas, the process becomes one of determining actual geographic locations having specific field strengths. Television broadcasters, for example, commonly determine their radiation pattern for a 100-mV/m field strength.

FREQUENCY AND WAVELENGTH

Frequency and wavelength are inversely related: that is, the higher the frequency, the shorter the wavelength; or, the lower the frequency, the longer the wavelength. The velocity with which radio waves travel through space depends on the characteristics of space just as the velocity with which a radio wave travels along a transmission line depends on the characteristics of the transmission line.

Space has two constants: one is the dielectric constant, which is 8.85 microfarads per meter; the other is the permeability constant, which is 1.257 microhenrys per meter. If you know the inductance and the capacitance, the velocity of propagation can be calculated as 300,000,000 meters per second, or approximately 186,000 miles per second.

The frequency of a radio wave is predetermined at the transmitter. When frequency is known, wavelength (λ) can be found by dividing the frequency into the velocity of propagation. The basic formula is:

$$\lambda = \frac{v}{f}$$

or,

$$\lambda = \frac{186{,}000 \text{ miles/second}}{f}$$

or,

$$\lambda = \frac{300{,}000{,}000 \text{ meters/second}}{f}$$

where,

f is in hertz.

If f is in megahertz, the formula is simpler:

$$\lambda = \frac{.186}{f}$$

or,

$$\lambda = \frac{300}{f}.$$

From this it can be seen that it is possible to calculate wavelength or frequency if only one of the two is known. Here are some examples. In each case, v is the velocity of propagation and f is the frequency in megahertz.

Let f = 100 MHz, then:

$$\lambda = \frac{.186}{100} = .00186 \text{ mile}$$

or,

$$\lambda = \frac{300}{100} = 3 \text{ meters.}$$

ANTENNA LENGTH

Although antennas of various lengths can be used for different applications, the half-wave antenna is used for most purposes. It is important to recognize, however, that a half-wave antenna is one-half wavelength long for one particular operating frequency only. When the operating frequency changes, the physical length of the antenna does not change, but its electrical length does change. This can be seen from the example given in Fig. 2-11, where, for purposes of explanation, the operating frequency is 300 MHz. The antenna is one-half wavelength long; one-half wavelength at this frequency is 50 centimeters (Fig. 2-11A). Considering that this antenna can be used at several different frequencies, it is important to observe what happens to the electrical length of the antenna at various operating frequencies.

In Fig. 2-11B the frequency has been increased to 600 MHz. At this frequency a half wavelength is 25 centimeters, but we are still using the 50-centimeter antenna; hence, at the new operating frequency of 600 MHz the electrical length of our antenna is now a full wavelength, whereas at 300 MHz it was one-half wavelength. Note that the physical length has not changed, but the electrical length has changed.

If the frequency is again doubled to 1200 MHz, a half wavelength becomes 12.5 centimeters; the same antenna that is 50 centimeters long physically will now be two full wavelengths long electrically, as

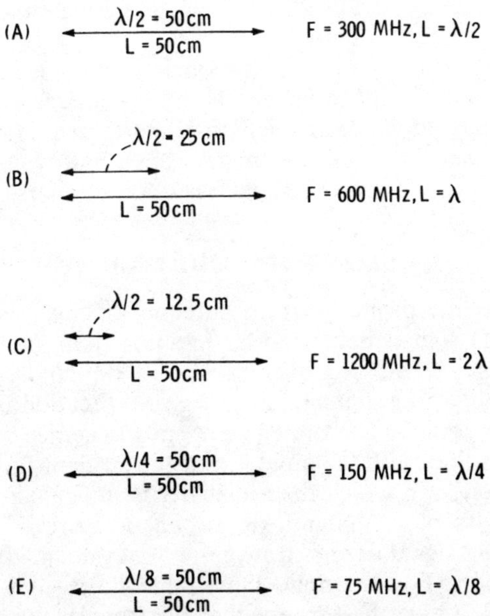

Fig. 2-11. Antenna length and frequency relationships.

shown in Fig. 2-11C. Thus, as the operating frequency is increased, the antenna becomes longer electrically without changing its physical dimensions. The opposite of this is true when the frequency is decreased. For example, at the frequency of 150 MHz (Fig. 2-11D), the 50-centimeter antenna becomes a quarter wavelength long electrically. In Fig. 2-11E the frequency is again reduced by one half to 75 MHz, making the 50-centimeter antenna one-eighth wavelength long electrically.

In the preceding examples, the physical length of the antenna was kept constant, but its electrical length changed with the frequency. It is also possible to cut an antenna to a half wavelength at any particular operating frequency. Thus, the physical length of the antenna is changed as the operating frequency is changed so that the electrical length stays constant. As stated earlier, at a frequency of 300 MHz, a half-wave antenna is 50 centimeters long. If the frequency is doubled to 600 MHz, the antenna must be half as long physically (25 centimeters) to be a half wavelength electrically. If the frequency is increased to 1200 MHz, the antenna must be reduced to 12.5 centimeters physically in order to remain a half wavelength electrically. Reducing the original 300-MHz frequency to 150 MHz means that the antenna must be increased to 100 centimeters long physically to be a half wavelength electrically. If the frequency is reduced to 75 MHz, the antenna must be made 200 centimeters long in order to be a half-wave antenna electrically.

Actually, the velocity of propagation is slightly less in a metallic conductor (a wire) than in space. Thus, the antenna length should be approximately 95% of the lengths determined in the preceding paragraph. Rather than 100 centimeters for a half-wave dipole operating at 150 MHz, the length should be 95 centimeters.

DIRECTIVITY AND GAIN

Directivity is the property of an antenna that causes it to receive (or to transmit) signals better in one direction than another. Directivity is most easily expressed by means of a directivity pattern. The relative sensitivity of an antenna in any given direction is indicated by the length of a straight line drawn between the center of the antenna and the pattern line, with the direction of the straight line corresponding to the direction in which the sensitivity is indicated.

The directivity of an antenna is also a measure of its ability to choose between signals from transmitters not located in the same angular direction. For example, the simple horizontal dipole is bidirectional and receives best from either the front or the back (at right angles to the antenna), as indicated by the shape of a figure-eight pattern. A unidirectional antenna is one that receives best in a single direction only.

Although a directional antenna receives best in its favored directions, it will also receive signals from other directions with less sensitivity. Directional patterns show how the sensitivity varies, not only for the favored directions, but also for other directions.

Antenna directivity patterns, like the antenna patterns discussed before, are displayed as shown in Fig. 2-12. This type of display is known as a *polar graph.* The antenna is assumed to be at the center of the graph, and the reception (or transmission) characteristics are plotted in a 360° rotation about the antenna. The shapes of the lobes of a pattern indicate the directional properties of an antenna. The pattern shown in Fig. 2-12 is for the standard, or reference, half-wave dipole and, as can be seen, this type of antenna is bidirectional; that is, it has equal lobes in both the forward and rear directions. The direction of maximum sensitivity is on a line directly perpendicular to the length of the antenna. Note that at the 70.7% point on the zero degree line, an arc has been drawn to one side of the pattern lobe. On a voltage polar pattern (Fig. 2-12), the 70.7% point is known as the half-power point. A similar half-power point exists on the opposite side of the lobe. The angle between the two half-power points is the *beam-width angle,* used to describe the relative directivity characteristics of various antennas. The standard dipole pattern of Fig. 2-12

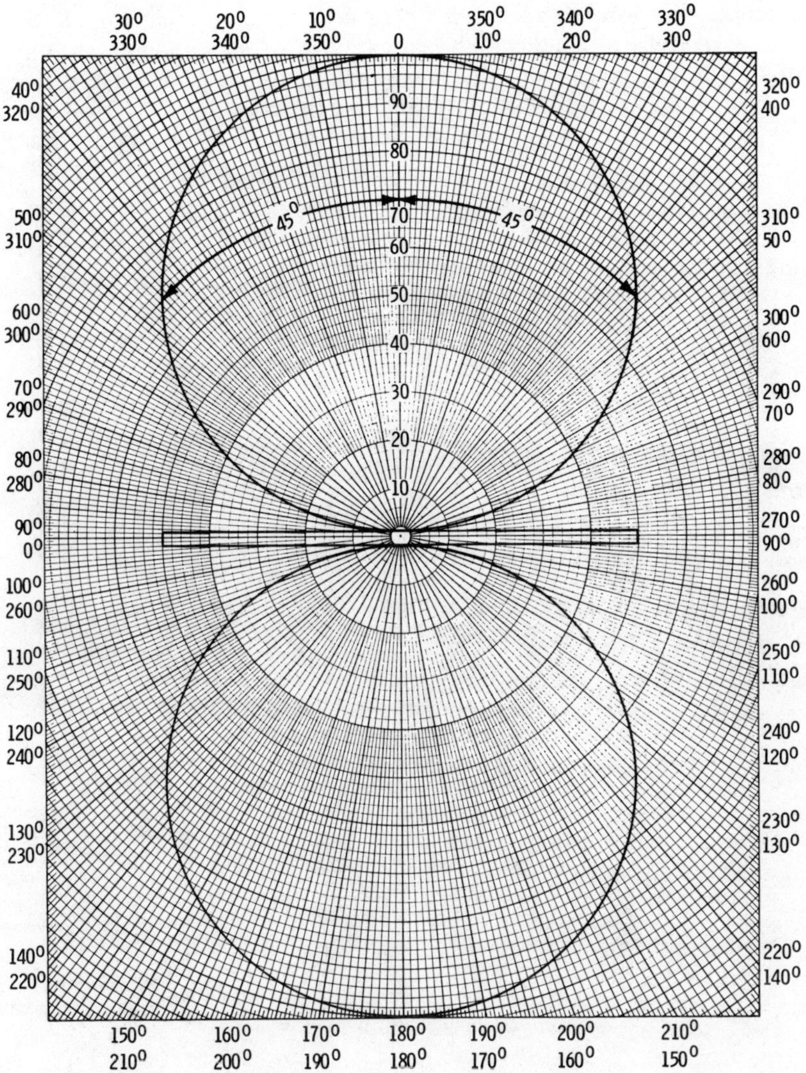

Fig. 2-12. Voltage polar pattern for standard half-wavelength dipole.

shows this angle to be 90°, which is also a standard or reference figure. In practical design work, values in the neighborhood of 8° have been attained by antenna manufacturers.

It should be noted also that there are directions of minimum reception in antenna directivity patterns between the lobes. These are called *nulls*. Often the nulls are more important than the lobes. For example, the lobes off the sides of a horizontal resonant dipole are broad; therefore, changing the orientation of the antenna through quite a few degrees from the maximum reception point of the lobe does not change the sensitivity any appreciable amount. However, off the ends, directly in line with the antenna, there are fairly sharp

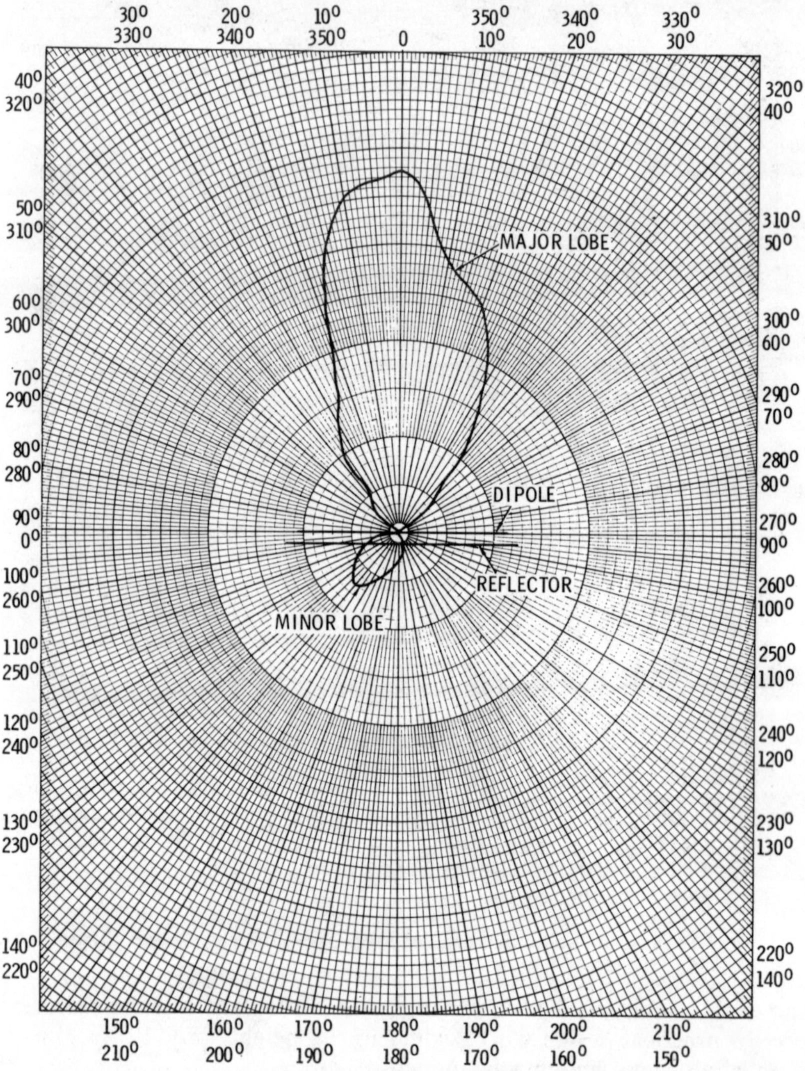

Fig. 2-13. Antenna pattern for a dipole with reflector showing major and minor lobes.

nulls, and in each null the reception is practically zero. In the case of two signals being received at the same frequency, it may be possible to eliminate one with the null portion while still receiving an adequate signal from the other source.

Somewhat in the same vein are the minor lobes of directional antennas, as seen in Figs. 2-13 and 2-14. The pattern of Fig. 2-13 is for a dipole with reflector, while the pattern of Fig. 2-14 is for a dipole with director. (Directors and reflectors are known as *parasitic* elements, which will be discussed in more detail in Chapter 3.) These minor lobes would appear to be insignificant. However, a nearby interference producer, such as a neon sign, within the angle of a minor lobe can completely wreck a television picture.

The gain of an antenna is a measure of the power induced by a constant signal in a given antenna as compared with that induced in a standard half-wave dipole cut to resonance for the exact frequency of the signal. A power gain of 10 means that the antenna in question is 10 times as effective as the standard dipole.* The *decibel,* which is an expression for power ratio, is often used, and the formula is:

$$\text{Decibels (dB)} = 10 \log_{10} \frac{P_2}{P_1}$$

Since the decibel is based on a ratio, the exact values do not necessarily affect the number of decibels. Rather, it is how one value compares with the other that determines decibels. If, for example, the gain of an antenna in decibels is given as 10, it means that the ratio (P_2/P_1) as defined in the foregoing is 10 to 1 because the log of ten is 1, and 10×1 is 10. In actual gain charts, the zero reference line at the bottom is taken as a gain of $1 = 0$ dB, or the actual power delivered by the dipole cut to resonance at this same frequency.

To understand the gain of an antenna, you must be familiar with the following terms:

Transmitting Gain—If an antenna radiates A watts and a standard dipole radiates B watts under the same conditions, the transmitting gain is A/B.

Receiving Gain—If an antenna receives A watts and a standard dipole receives B watts under the same conditions, the receiving gain is A/B.

* Some effort is being made to relate antenna gain to the hypothetical *isotropic antenna,* which would radiate equally in all directions. While such an antenna is impossible, its conception is a basis for agreement and understanding that seems logical.

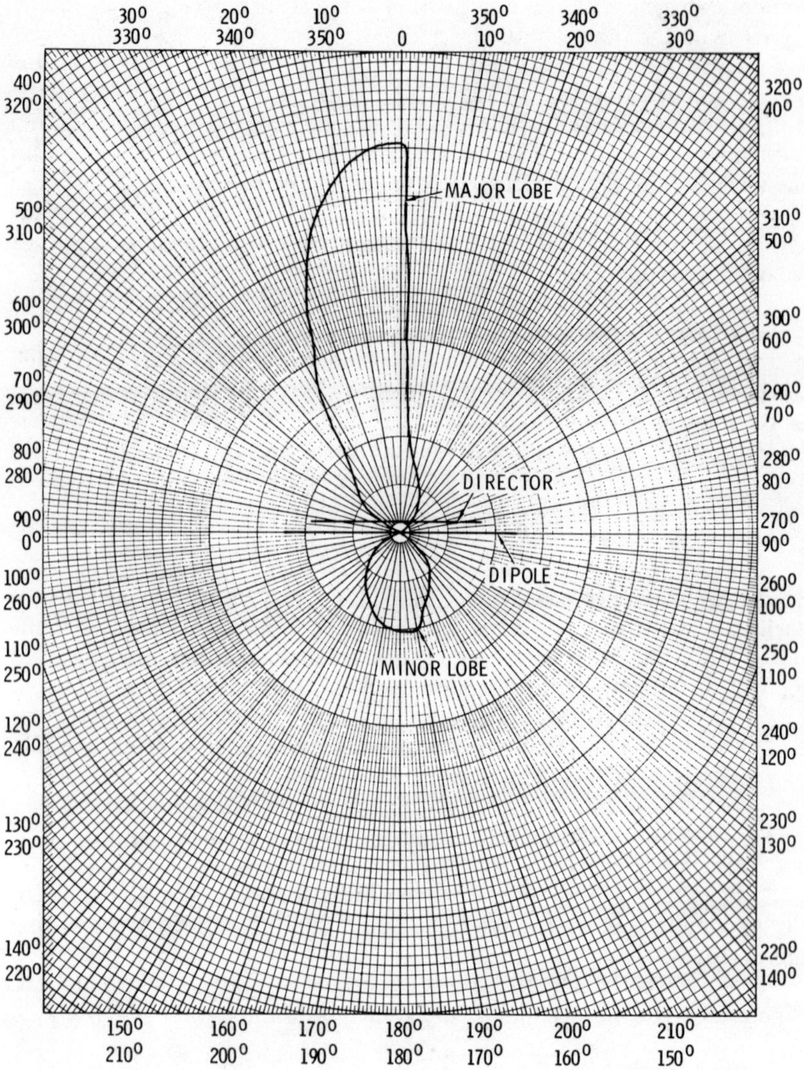

Fig. 2-14. Antenna pattern for a dipole with director showing major and minor lobes.

Standard Dipole—A standard half-wave antenna (dipole) is considered to have unity gain (0 dB), and other antenna gains use this as a reference.

Erp (*Effective Radiated Power*)—The gain of an antenna multiplies the output, so an antenna with a gain of G and a power input to the antenna of M watts would have an erp of G × M.

EXAMPLES:

1. A transmitting antenna radiates 1000 watts; a standard dipole radiates 500 watts. What is the transmitting gain?

$$\text{Gain} = \frac{1000}{500} = 2$$

2. A receiving antenna receives 0.5 watt, and a standard dipole receives 0.3 watt. What is the receiving gain?

$$\text{Gain} = \frac{0.5}{0.3} = 1.7$$

3. A transmitting antenna has a gain of 10 and a power input of 600 watts. What is the effective radiated power (erp)?

$$\begin{aligned}\text{Erp} &= \text{gain} \times \text{power input}\\ &= 10 \times 600 = 6000 \text{ watts}\end{aligned}$$

3

Basic Antenna Designs

This chapter covers certain basic antennas including horizontal and vertical types, as well as antenna arrays and the use of parasitic elements. Since there are so many different types of antennas covering a wide variety of applications, some special types are also covered.

BASIC ANTENNA TYPES

Horizontal Antennas

A fundamental horizontal antenna is the half-wave dipole shown in Fig. 3-1A. The television receiving antenna is probably the most widely used horizontal type.

The horizontal half-wave antenna is a special type of tuned resonant circuit, and the Q of the antenna, like the Q of any resonant circuit, depends on the relation of reactance to resistance. The response curve of a half-wave antenna is very similar to that of an ordinary low-frequency inductance and capacitance resonant circuit. The sharpness of the response curve depends on the construction used in the antenna. When made from a thin copper wire, the antenna will

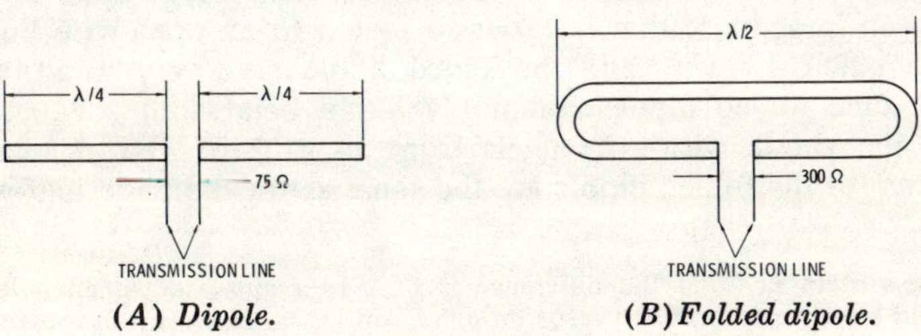

(A) Dipole. *(B) Folded dipole.*

Fig. 3-1. Basic horizontal antennas.

act like a high Q or very sharply tuned circuit. Such an antenna will operate only over a narrow band of frequencies, the same as any sharply tuned resonant circuit. As the diameter of the conductors is increased in relation to the overall length, the antenna becomes less selective in its tuning and will cover a greater band of frequencies.

In electrical terms, the dipole or half-wave antenna is the equivalent of an open-circuited quarter-wave transmission line. Because of its resonant characteristics, the half-wave dipole builds up larger voltage and current standing waves and, in turn, radiates considerable energy. The standing waves also define the physical wavelength of the antenna. Because of the resistance of the wire, however, the movement of radio waves along the antenna is somewhat slower than the movement of radio waves in space. Hence, the wavelength of an antenna is slightly less than that of a free space radio wave. A dipole antenna is about 5% shorter than the theoretical half wave in space* For example, at 300 MHz a full wavelength is 100 cm. The dipole cut for this frequency should be 47.5 cm [½ (95% × 100)] to compensate for the 5% difference.

The basic dipole antenna, fed at its midpoint, has an impedance of about 75 ohms. While amateur radio operators and electronic engineers would have little trouble matching this impedance to any transmission line, such impedance matching becomes too involved for mass installations such as television. That is, even though the television receiving antenna is installed by a qualified technician, the technician cannot waste time with impedance matching. Coaxial cable, which does match the 75-ohm dipole, is too costly. Open-wire transmission lines must be built at the site and have characteristic impedances of 400 ohms or higher. There is a 75-ohm "twin lead" available, but its losses are too great to be satisfactory. The answer was a variation of the half-wave dipole known as the *folded dipole*.

The folded dipole is, in effect, a single half-wave conductor placed above the two quarter-wave conductors of a simple dipole, with the ends connected together as shown in Fig. 3-1B. The voltage induced into this arrangement is four times that induced into the basic dipole, but the output current is the same. Thus, the folded-dipole impedance is four times that of a simple dipole, or 300 ohms. In addition to providing an impedance that is easier to match to an open-wire line or a fairly efficient "twin lead," the folded dipole has a very broad bandwidth. One folded dipole commonly has a bandwidth adequate to cover the vhf television channels from 54 to 216 MHz. Radiation patterns for the folded dipole are the same as for the basic half-wave

* Some writers say that the difference is 6%. In reality, the antenna length should be adjusted until the resonant condition exists. However, in most situations, the 95% (100% − 5%) factor will give a length that is satisfactory.

dipole. Such bidirectional patterns can have disadvantages as well as advantages. In situations where nondirectional reception or radiation is desired, it becomes necessary to use two dipoles, or folded dipoles, crossed in a turnstile arrangement as shown in Fig. 3-2. For unidirectional radiation, the use of driven or parasitic arrays, discussed later in this chapter, is required.

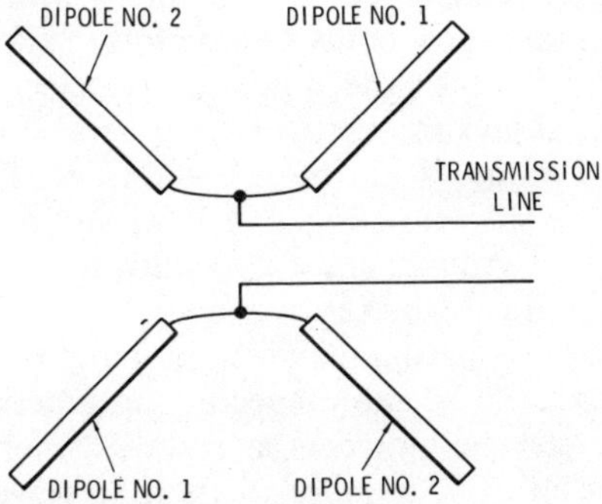

Fig. 3-2. Dipoles connected in a turnstile arrangement to provide an omnidirectional pattern.

Vertical Antennas

The vertical antenna deserves particular consideration because of its widespread use. With an omnidirectional radiation pattern, the vertical antenna is used in most mobile communication services—mobile communications being an integral part of our highly mobile society. Its omnidirectional (or nondirectional) pattern provides an overall coverage within a given area. For this reason, the police radio at or near the center of a city can communicate with any patrol car within the city. The vertical antenna also has the advantage of requiring little mounting space.

The fundamental vertical antenna is the grounded quarter-wave radiator historically known as the *Marconi* antenna (Fig. 3-3A). The vertical portion above ground is one-quarter wavelength, and there is a reflected or "image" antenna in the ground itself, thus forming an effective vertical half-wave dipole. The vertical radiation pattern of such an antenna is similar to half a doughnut, with the antenna sticking up through the hole in the doughnut (Fig. 3-3B). Thus, the horizontal radiation pattern of a vertical antenna is circular or omnidirectional (Fig. 3-3C).

The actual length of a vertical antenna ranges from one-quarter wavelength to as much as two wavelengths, depending on the coverage desired and the space available. As we shall see later, the length is relative to sky-wave reduction and sensitivity. The impedance, or radiation resistance, of a vertical antenna is usually somewhere about the 30-ohm level, depending greatly on the grounding system. To match this 30-ohm impedance to a 75-ohm transmitter or receiver, a length of 50-ohm coaxial cable is used. As indicated by Fig. 3-4A, the vertical antenna is commonly mounted on a mast or pole well above the actual ground level. However, this puts the actual antenna and its reflected image beneath the ground level too far apart and increases the ground reseistance. By placing a number of horizontal (or nearly horizontal) grounded conductors (known as a *ground plane*) immediately beneath the antenna, the electrical ground level is raised. Fig. 3-4A shows the vertical antenna and its reflected image without the ground plane. With the ground plane, the actual antenna and the image are as shown in Fig. 3-4B. In addition to reducing the ground resistance, the ground plane concentrates the radiations along the horizontal and reduces the sky wave.

A vertical antenna is resonant when its length is any multiple of a quarter wavelength. To be more specific, an antenna having an odd number of quarter wavelengths acts as a series-resonant circuit, while an antenna having an even number of quarter wavelengths acts as a parallel-resonant circuit. Antennas that are shorter than these quarter wavelengths are capacitive in nature; those that are longer are induc-

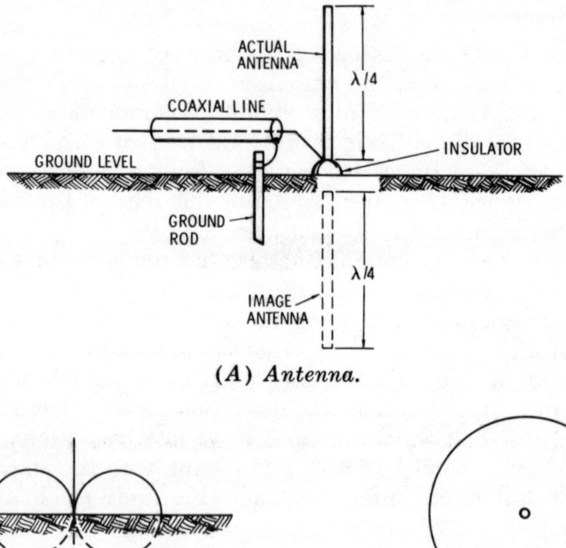

(A) Antenna.

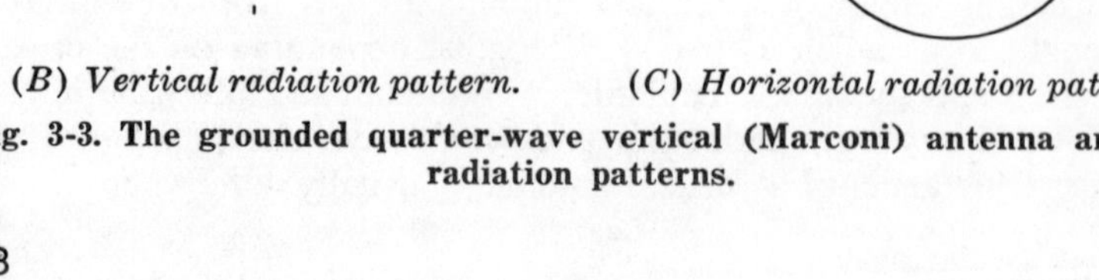

(B) Vertical radiation pattern. *(C) Horizontal radiation pattern.*

Fig. 3-3. The grounded quarter-wave vertical (Marconi) antenna and its radiation patterns.

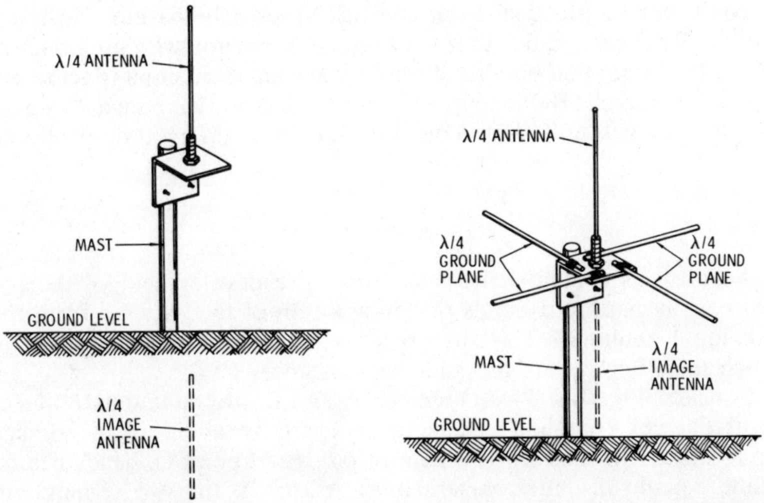

(*A*) *Without ground plane.* (*B*) *With ground plane.*

Fig. 3-4. Quarter-wave vertical antennas with and without ground plane.

tive in nature. The addition of an actual inductor at the base or at the center of an antenna reduces its capacitance and gives the antenna a longer electrical length. For example, to achieve the optimum ground-wave coverage in mobile amateur radio, an antenna of ⅝ wavelength is preferred. On the popular 10-meter (28-MHz) amateur band, the ⅝ wavelength measures 22+ feet (6.75 meters). Obviously,

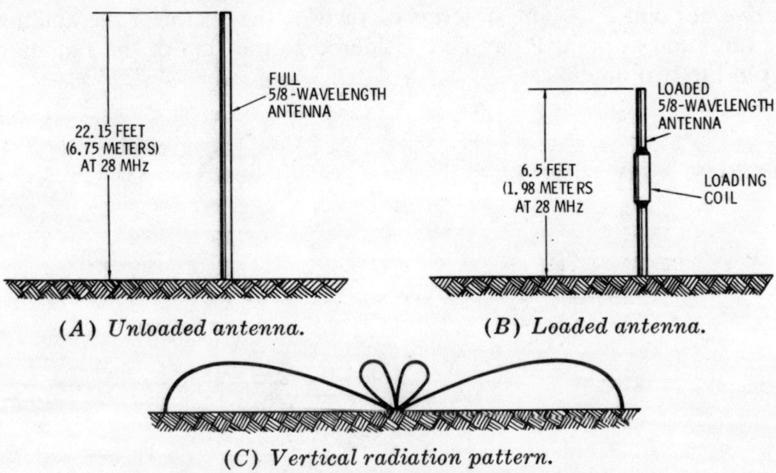

(*A*) *Unloaded antenna.* (*B*) *Loaded antenna.*

(*C*) *Vertical radiation pattern.*

Fig. 3-5. Comparison of lengths for loaded and unloaded 5/8-wavelength antennas, and the vertical radiation pattern for both.

it would not be practical to mount a 22-foot antenna atop an automobile. By *loading* a 6.5-foot (1.98-meter) antenna with an inductor, the vertical radiation pattern of the ⅝-wavelength antenna is achieved. Fig. 3-5 compares the lengths of these loaded and unloaded ⅝-wavelength antennas and shows the vertical radiation pattern produced by each.

ARRAYS

An array is a multielement antenna. The development of the antenna array pattern depends on the addition of the patterns from the individual antennas. This in turn depends on the relative phase in which the antennas are fed and their distances apart.

Consider Fig. 3-6. Two antennas, A and B, are mounted 180 electrical degrees (one-half wavelength) apart. In addition, the signals fed to these antennas are 180° out of phase. At point Q, which lies on a line equally distant from antennas A and B, the two signals will cancel because their path lengths are exactly the same and the signals are exactly out of phase. This is also shown in the vector diagram; the two signals at point Q are equal but opposite in phase, and the vector resultant is zero.

If, however, we consider a point P, which lies on a straight line drawn through antennas A and B, there will be signal addition. Since one path is from antenna B to point P, and the second path is from antenna A to point P, the two path lengths differ by 180 electrical degrees. And, since the two signals left the two antennas in an out-of-phase condition, the radiation patterns will add when the signals arrive at point P. Again, in terms of vectors, the vector of the amount of radiation at point P can be considered as the sum of the radiation from the two antennas.

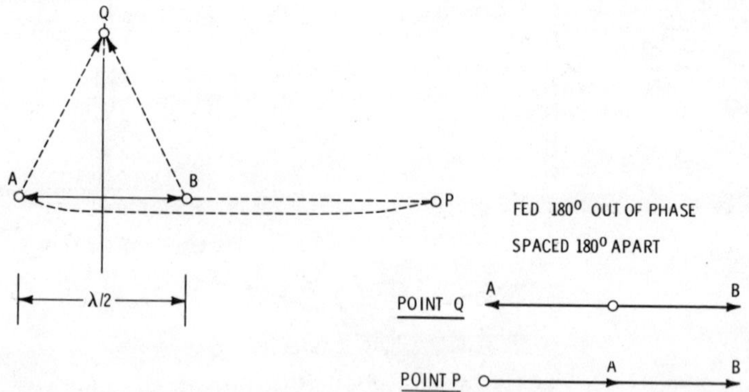

Fig. 3-6. Development of antenna array pattern. Antennas mounted 180 electrical degrees (1/2 wavelength) apart; fed 180° out of phase.

Thus, we can see from a simple vector analysis that there will be an addition of signal strength on a line drawn through the antennas, and a cancellation of signal strength on a line perpendicular to the line between the antennas.

Consider another case, diagrammed in Fig. 3-7. Here we have two antennas 90 electrical degrees (one-quarter wavelength) apart but fed in phase. When we examine the two vectors as before, we can see that at point Q there will be signal addition, since the two individual signals are in phase and the path lengths from Q to A and from Q to B are exactly equal. Thus, at point Q we will have a direct addition of the two signals. However, at point P the signal from antenna B has the distance BP to travel, while the signal from antenna A travels the distance AP, or 90° farther. The two signals will not cancel at point P, nor will they add directly. Rather, they will add in a vectorial fashion so that the amount of radiation at point P will be greater than for a single antenna but less than for two antennas. This is shown in the vector diagram.

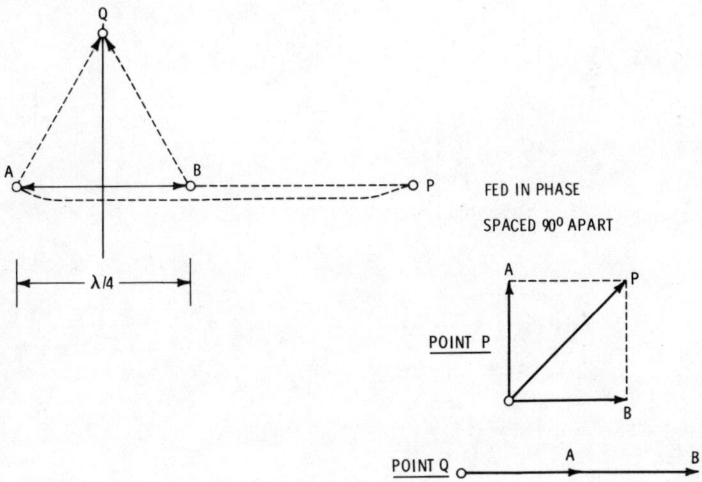

Fig. 3-7. Development of antenna array pattern. Antennas mounted 90 electrical degrees (1/4 wavelength) apart and fed in phase.

End-Fire Array

While there are many useful combinations of antenna patterns, two of the more fundamental types are shown in Fig. 3-8. Again we are looking down on vertical antennas so that we see their horizontal radiation patterns. Fig. 3-8A is referred to as an "end-fire" pattern. In this type of pattern, the two antennas are 180 electrical degrees (one-half wavelength) apart and are fed 180° out of phase. As discussed earlier, this causes a signal addition on a line drawn through the antennas, and a signal cancellation at any point equidistant from the antennas (i.e., on a line drawn perpendicular to the line A-B at the exact midpoint between the antennas). Thus, if we consider antennas A and B a transmitter, with four receivers located at points P, Q, R, and S, there will be maximum signal at receivers P and S, and minimum signal at receivers Q and R. Note that a minimum signal is not the same as a zero signal because there will be some signal at points Q and R, althought it will be small.

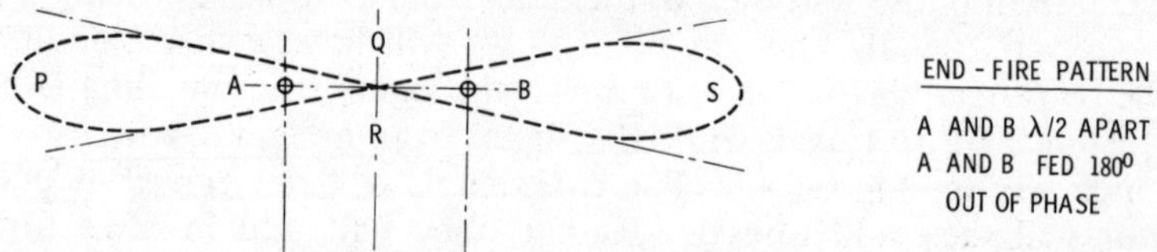

(A) "End-fire" pattern.

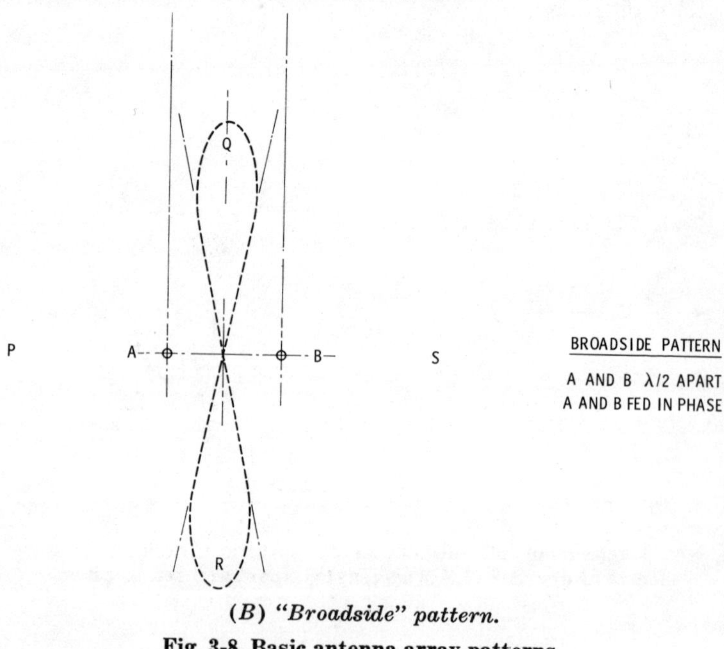

(B) "Broadside" pattern.

Fig. 3-8. Basic antenna array patterns.

Broadside Array

Fig. 3-8B shows a "broadside" antenna pattern. Here the antennas are again one-half wavelength apart but are fed in phase. There will be maximum signal at receivers Q and R, and minimum signal at receivers P and S. Thus, it is possible, in this rudimentary system, to direct the antenna pattern in a given direction—either in line with the antennas, as in Fig. 3-8A, or at right angles to this line, as in Fig. 3-8B.

The pattern can be made sharper by increasing the number of antennas. This will extend and lengthen the beam. Consider Fig. 3-9. Here there are four antennas—A, B, C, and E—with each pair a half wavelength apart. Again, the antennas can be fed in one of two ways. If they are fed out of phase, there will be maximum signal at point P, which is on a straight line connecting the antennas, and there will be minimum signal at point Q, which is at right angles to a line from A to E. If the antennas are fed in phase, there will be maximum signal at point Q and minimum signal at point P.

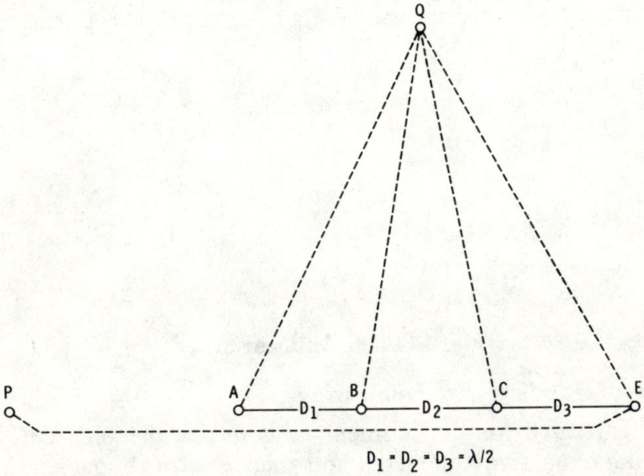

Fig. 3-9. Increasing the number of antennas sharpens the array pattern.

Collinear Array

Another basic type of directional antenna is the collinear array shown in Fig. 3-10. The collinear is a version of the broadside array, having a number of half-wavelength (or ⅝-wavelength) elements stacked end-to-end and fed in phase. Quarter-wave phasing stubs are connected between the elements to provide the necessary phase reversal every half wavelength, thus maintaining an in-phase current condition in each radiating element. The array is usually fed at the center so that the energy is distributed evenly among the radiating elements.

Collinear arrays may be mounted vertically or horizontally. When mounted vertically, the collinear provides the same omnidirectional horizontal pattern as a single vertical antenna, but the radiation is concentrated at low vertical angles similar to the pattern of Fig. 3-5C.

When mounted horizontally, the vertical radiation of the collinear is the same as for a single antenna, but the horizontal pattern is extended and made more sharply directional as in the broadside array pattern of Fig. 3-8B.

Modern versions of the vertical collinear antenna have the radiating elements and phasing circuits encapsulated within a fiber-glass case. Used primarily for base-station installations, these vertical collinears are said to have gains up to 9.5 dB and a vertical beam width of 8°.

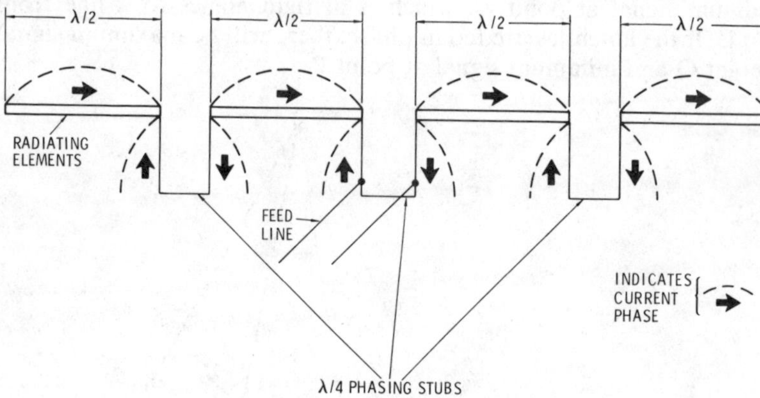

Fig. 3-10. The collinear array.

Methods of Feeding Antenna Arrays

It is instructive for the moment to consider the physical spacing between antenna array elements and their electrical spacing. Various techniques are used to feed the array elements at almost any desired phase relationship and with almost any desired physical spacing.

Fig. 3-11A shows antennas A and B spaced one-half wavelength apart and fed in phase with an open-wire line. The line from the transmitter (or receiver) connects at the midpoint of the line between A and B. Thus, the distance from the transmitter to antenna A is exactly equal to that distance from the transmitter to antenna B. Electrically, the transmitter is equally distant from A and B and the two antennas are fed in phase. A similar situation, using coaxial cable, is shown in Fig. 3-11B.

In Fig. 3-12A, antenna A is fed one-half wavelength (180 electrical degrees) out of phase with the power fed to antenna B. The cable length between junction J and A is made equal to a full half wavelength, while that between J and B is a half wavelength multiplied by the factor (1 – VF). (VF is the velocity-of-propagation factor of the transmission line.) For an open-air line having a VF of 95%, the distance JB becomes insignificant and has an appearance somewhat

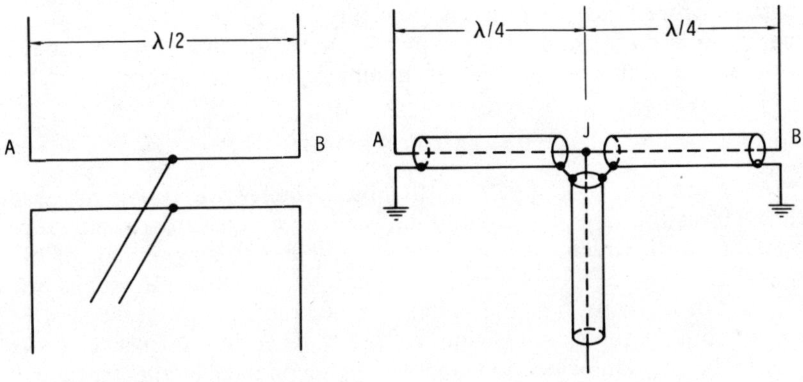

(*A*) *With open-wire line.* (*B*) *With coaxial line.*

Fig. 3-11. Methods of feeding antennas in phase.

like that of Fig. 3-12B. However, a VF of 66%, for some types of coaxial cable, would give JB a physical length of 0.5 (1 − 0.66) or 0.17 wavelength. Electrically, the length of JA is 0.757 wavelength (0.5 ÷ 0.66) and JB is 0.257 wavelength. This means that in traveling from the transmitter to antenna A, the wave travels 0.5 wavelength farther than that traveling to antenna B. Thus, the power fed to antenna A is one-half wavelength (180 electrical degrees) out of phase with that being fed to antenna B.

To develop a phase difference of 90°, the cable JB can be given an electrical length equivalent to 0.507 wavelength. If JA has an electrical equivalent of 0.757 wavelength, the difference between JA and JB produces the 90° phase shift. Of course, neither this nor any of the preceding can account for impedance matching. Two or more antennas with their connecting cables joined at a junction, J, give an

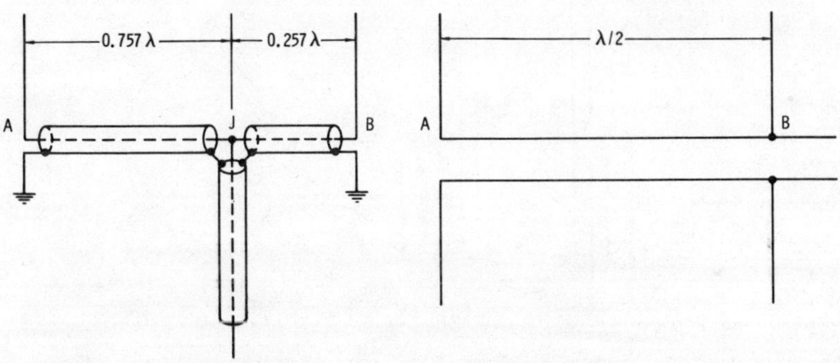

(*A*) *With coaxial line.* (*B*) *With open-wire line.*

Fig. 3-12. Methods of feeding antennas 180° out of phase.

effective impedance that is less than any of the branches. So, while matching is mathematically beyond the scope of this book, we have covered the primary objective of antennas and their arrays.

PARASITIC ELEMENTS

As discussed in Chapter 2, the basic half-wave dipole is used throughout the communications industry as a standard or reference antenna with which the characteristics (particularly gain) of all other antennas are compared. The gain of an antenna means the amount of increase in signal voltage induced in the driven element of the antenna in question, over the voltage induced in a reference dipole placed in the same location and cut to resonance at the frequency being used. Antenna gain is always expressed as a voltage ratio and

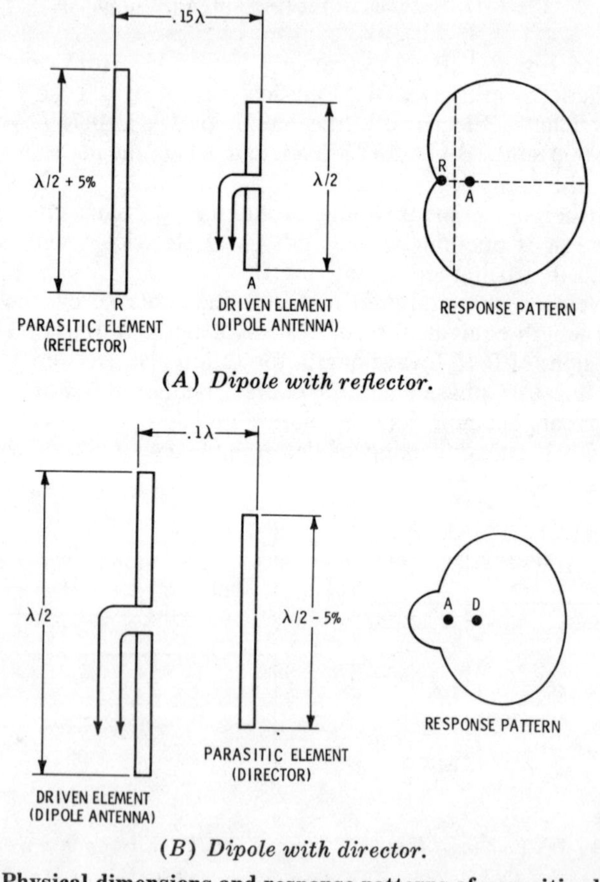

(A) Dipole with reflector.

(B) Dipole with director.

Fig. 3-13. Physical dimensions and response patterns of parasitic elements.

is stated in terms of decibels. Thus, if a certain antenna develops twice as much signal voltage as a reference dipole, the first antenna is said to have a gain of 6 dB.

The gain of an ordinary dipole can be increased by the addition of parasitic elements positioned either in front of or behind the dipole or driven element. These parasitic elements are called *directors* or *reflectors* depending on their length and positioning with respect to the driven element. Although they are not connected directly to the driven dipole element, signal energy induced in them is coupled to the dipole through induction and radiation fields. Fig. 3-13 shows the spacing and resultant polar response patterns for the two types of parasitic elements.

Reflector

When a parasitic element is positioned approximately 0.15 wavelength behind the driven element (with respect to the direction from which a signal is being transmitted) and is made approximately 5% longer than the driven element, it is referred to as a *reflector*. In Fig. 3-14, the effect of a reflector on a passing radiation field is illustrated.

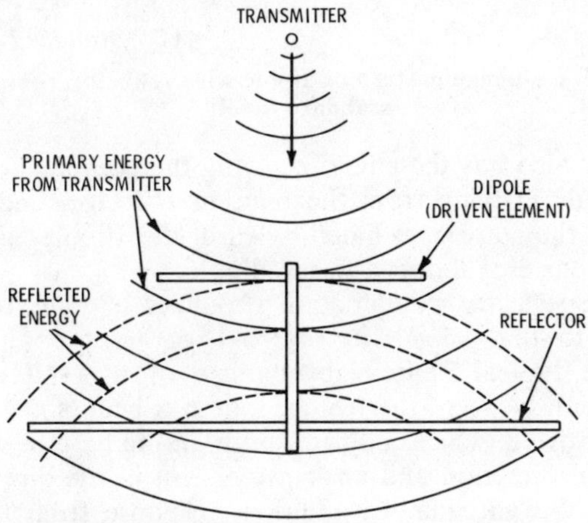

Fig. 3-14. Effect of reflector on radiation field.

The wavefronts of the passing field are shown cutting first through the driven dipole and then through the reflector. These wavefronts are shown as solid lines. The dotted-line waves indicate the action of the reflector. Signal currents are induced in the reflector just as they are in the dipole, but since the reflector is not connected to a resistive load, most of the energy contained in the reflector currents is reradiated. Because of the spacing between the reflector and the dipole, the reradiated, or "secondary," energy from the reflector arrives back at the dipole in phase with the primary energy from the transmitter. The reflected energy thus induces currents in the dipole, and these currents add to those induced by the primary signal. This antenna combination therefore exhibits *gain,* or an increase in signal pickup over that which would be picked up by the dipole element alone.

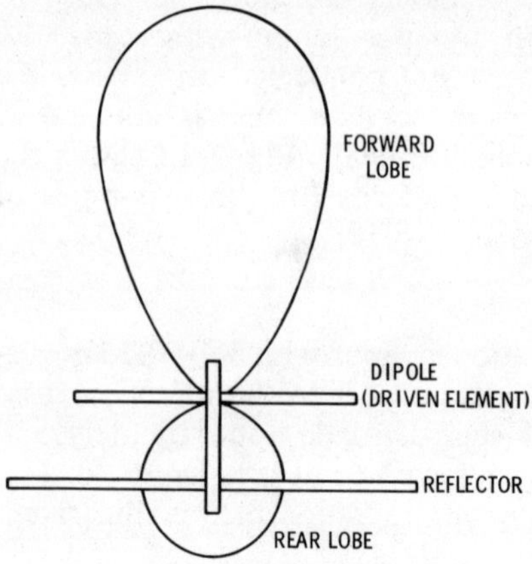

(*A*) *Dipole with reflector.*

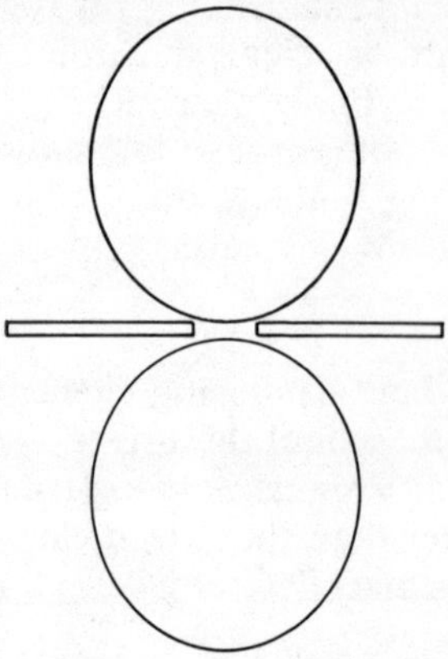

(*B*) *Standard dipole.*

Fig. 3-15. Response pattern of dipole with reflector, compared to standard dipole.

A reflector also has the effect of giving the combination directivity because signals arriving from the rear, or reflector, end will be out of phase as they are reradiated toward the dipole and will thus decrease the currents induced in the dipole.

The dipole-reflector combination produces a polar response pattern similar to that shown in Fig. 3-15A. (For comparison, Fig. 3-15B is the typical figure-eight pattern of a standard half-wave dipole.) As can be seen, the frontal lobe has been somewhat lengthened and narrowed, which indicates both an increase in signal pickup in the forward direction and an improvement in the directional characteristics of the antenna. In addition, response from the back has been considerably reduced, as shown by the smaller rear lobe.

The amount of signal pickup from the front compared with that from the rear provides us with another means of rating antennas. Known as the "front-to-back ratio," it is an important consideration in many types of radio-wave communication.

Director

The other type of parasitic element is the *director* (Fig. 3-13B). The director is usually spaced about 0.1 wavelength in front of the driven element and is usually about 5% shorter than the driven element. It aids in building up signal currents in the driven element in much the same manner as the reflector. Its major effect, however, is a narrowing of the polar response pattern in the forward direction. The addition of more directors increases the gain only slightly but narrows the frontal lobe considerably, making the array more and more directive. Fig. 3-16 illustrates the difference in the polar pat-

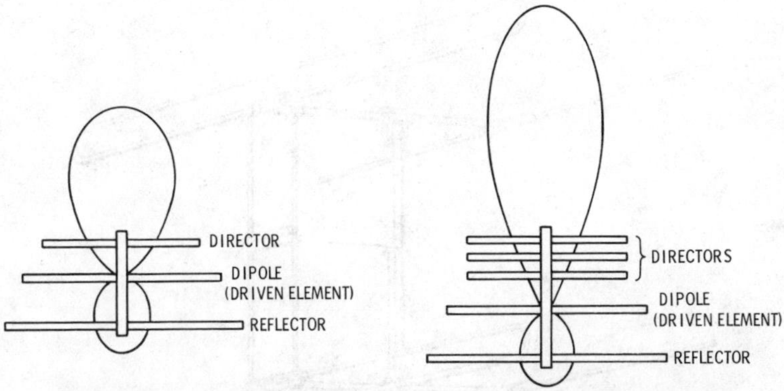

(A) One director. *(B) Three directors.*

Fig. 3-16. Effect of additional directors on response pattern.

terns of an array using one director and another array using three directors.

STACKED ARRAYS

Although both driven and parasitic arrays alter the horizontal radiation patterns, the vertical patterns remain much the same as they are in the horizontal dipole. Arrays that reduce radiation from bidirectional to almost unidirectional do eliminate some of the vertical "backside" radiation. However, such arrays do very little toward reducing "frontside" vertical radiation. And since most of these arrays are not intended for sky-wave propagation, vertical radiation is undesirable. One means of reducing vertical radiation is by stacking two or more antennas and their related parasitic elements vertically, one above the other.

Fig. 3-17A shows two folded dipole antennas with parasitic reflectors in a stacked arrangement. The two folded dipoles are fed in-phase and vertically spaced a quarter wavelength apart. Referring to the response pattern shown in Fig. 3-13A, a front-to-back ratio of about 4 to 1 is indicated. In the stacked arrangement, the two patterns add directly in the forward and backward directions but the addition becomes vectorial in the up and down directions. While the resultant response pattern is similar in shape to the single dipole/reflector combination, the elongated and slightly narrower front lobe indicates that much less of the radiation is going skyward (Fig. 3-17B). Or, if a stacked arrangement is being used for reception, the response to any signals coming from the sky is reduced. Still further reduction in sky-wave reception or radiation can be accomplished by vertically spacing

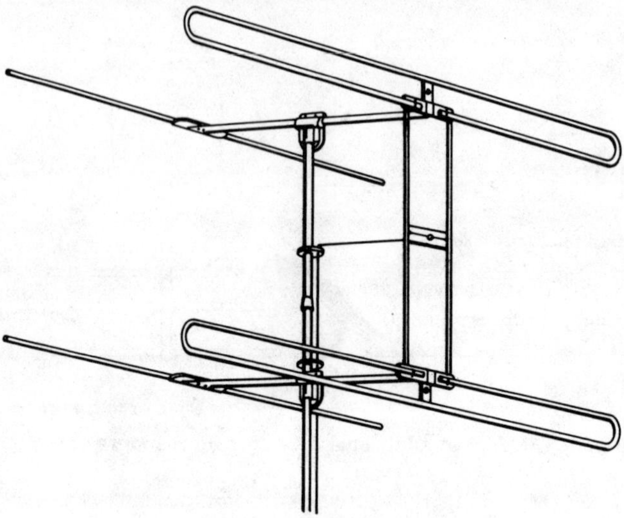

(A) Stacked folded dipoles with reflectors.

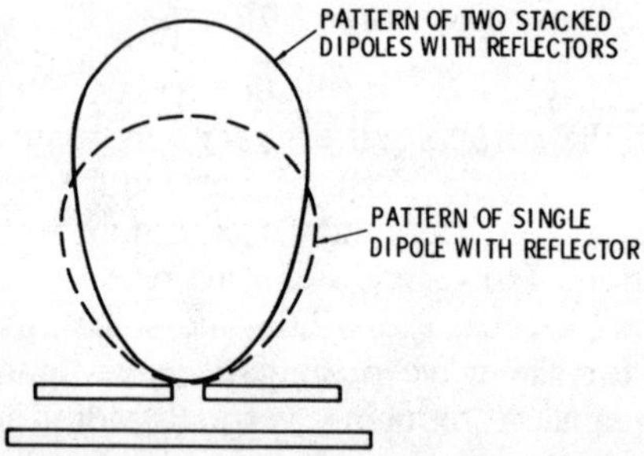

(B) Horizontal radiation patterns (backside radiation not shown).

Fig. 3-17. Effect of stacked antenna arrays on horizontal radiation pattern.

the dipoles a half wavelength apart or by stacking additional elements. While there is usually a physical limit to the number of elements that can be stacked for television reception and for amateur radio, commercial fm and television broadcasters stack as many as 12 crossed dipoles on a tower as a means of acquiring the desired power gain.

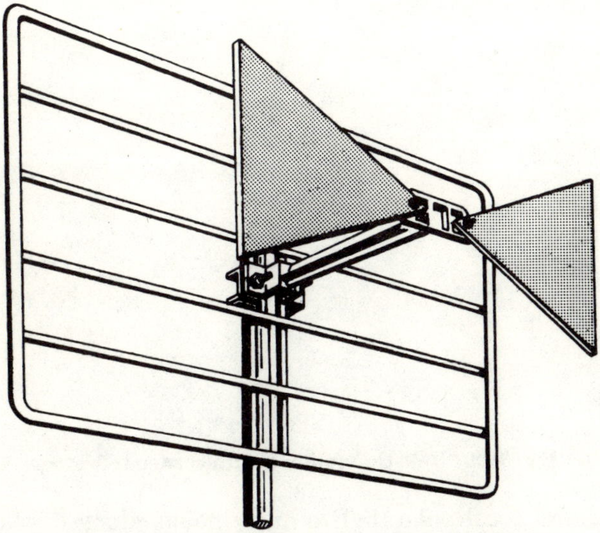

Fig. 3-18. Uhf "bow-tie" antenna with reflector.

REFLECTORS

As we have seen, directional antenna systems can be produced by using several antenna elements spaced and phased in such a manner that the basic radiation pattern is altered. Another method is to use reflectors in conjunction with a single active dipole element. In uhf television, for example, a reflector is often used as in Fig. 3-18. Here the active element is a "bow tie" (two triangular elements arranged to form a dipole). Elements of this type have a greater bandwidth than thin-wire elements.

Some signal energy arriving from the front will be captured by the dipole, while some will strike the reflector elements and "bounce" back to the dipole. Because of the spacing between the reflector and the dipole, the reflected energy will add to the energy received by the dipole. Energy arriving from the back will not be picked up by the dipole, making this a unidirectional antenna system. An open structure is used for the reflector since it has less wind resistance than a solid reflector.

Greater gain is possible through the use of a corner reflector like the one shown in Fig. 3-19. Each leg of the reflector system has 12 grounded dipoles. Their combination is very effective in providing a sharp forward beam.

Fig. 3-20 shows four radio waves (A, B, C, and D) entering a corner reflector. As the drawing illustrates, the waves are not reflected

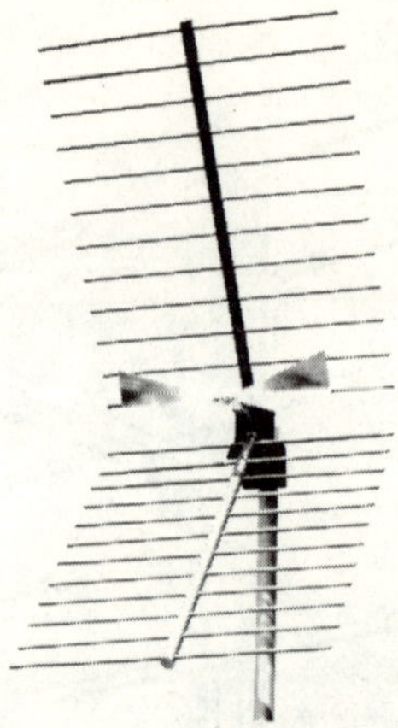

Fig. 3-19. "Bow-tie" antenna with corner reflector.

to one central focal point, but to many points along the line represented by points H and G. Thus, there is gain for this antenna, but since there is no single focal point the gain is less than for a parabolic reflector.

Parabolic Reflectors

Because of the quasi-optical properties of microwaves (frequencies above 1000 MHz), transmitting antennas are designed to focus their radiation into a concentrated beam of energy. Standard radiation patterns and methods of producing such patterns used at lower frequencies are not practical at microwave frequencies. The reason is the rapidity with which ground-wave propagation is attenuated at these extremely short wavelengths. Sky-wave propagation is nonexistent at these frequencies because microwaves are not affected by the ionosphere; instead, they pass on through to outer space. Transmission, therefore, is limited to direct, point-to-point space-wave propagation. To achieve maximum possible energy transfer between transmitting and receiving stations at practical distances (20 to 40 miles), para-

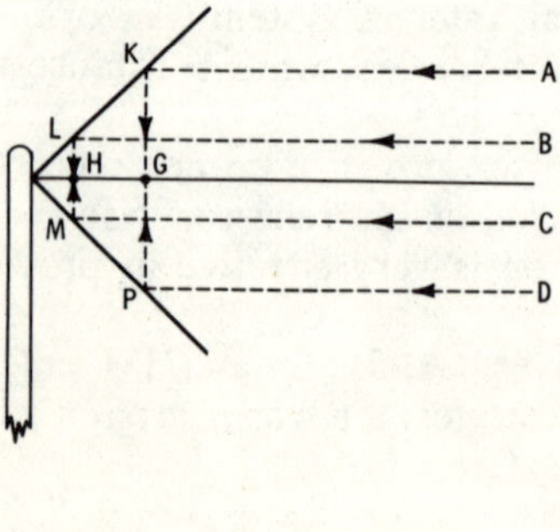

Fig. 3-20. Wave action in corner reflector.

bolic reflectors are used. They concentrate the radiated energy into a narrow beam, just as a flashlight reflector (also parabolic in shape) focuses light rays from a small electric bulb into a beam of light. Several types of parabolic reflectors are shown in Fig. 3-21.

The foregoing refers primarily to ground relay stations where various types of signals (television, telephone, police radio, power control, etc.) are transported over large distances via microwave beams rather

(*A*) *Dipole and reflector feed system.*

(*B*) *Waveguide feed system.*

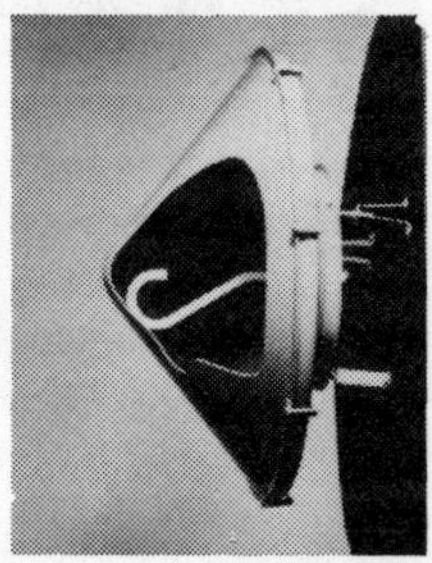

(*C*) *Waveguide feed inside weather protective covering* (*radome*).

Courtesy Jerrold Electronics Corp.

Fig. 3-21. Types of parabolic antennas.

than elaborate networks of cables. Coast-to-coast network television, for example, owes its existence to this efficient method of interconnecting commercial stations over long distances. Fig. 3-22 shows the parabolic "dish" antennas atop the tower of a typical microwave relay station.

The parabolic shape is used for a reflector at microwave frequencies because of its fundamental characteristic of focusing incoming rays or radio waves into a single point, or *focus.* This is illustrated in Fig. 3-23A. It is basic that the mathematically derived curve known

Fig. 3-22. Parabolic antennas on microwave relay tower.

as a parabola* will reflect any ray or wave from its surface to the focus point, regardless of the angle of incidence. Conversely, when a source of radiation is placed at the focus point, energy will be reflected from the parabola as a concentrated beam.

As shown in Fig. 3-23B, each wave follows the law of reflection in which wave CO strikes the surface at O and is reflected as wave OD. The angle of incidence (ϕ) is equal to the angle of reflection (θ). Each of the waves in Fig. 3-23A behaves just like the single wave in Fig. 3-23B. It is the nature of a parabolically shaped surface that all entering waves will be reflected to the focus point regardless of their angle of approach.

* The usual definition of a parabola is "the path of a moving point that remains equally distant from a fixed point (focus) and from a fixed straight line (directrix)."

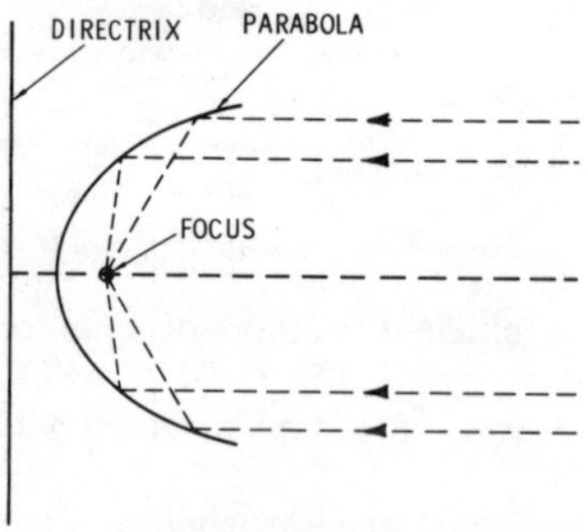

(A) Wave action in parabolic reflector.

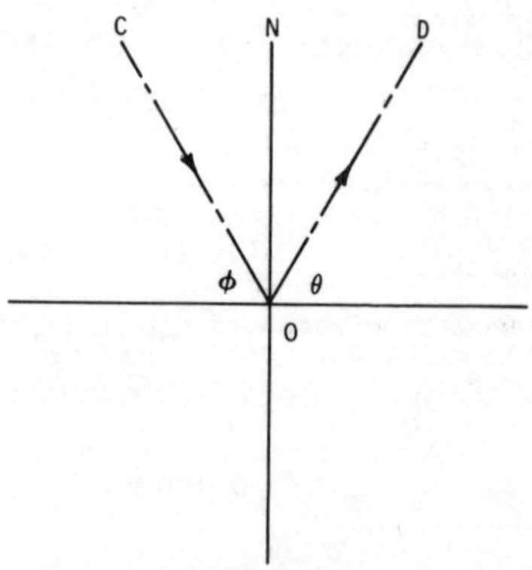

(B) Diagram illustrating law of reflection.

Fig. 3-23. A parabolic reflector focuses incoming waves to a single point.

SPECIAL ANTENNA TYPES

Antennas cover such a wide range of applications that an entire book could be devoted to the various uses. Some of these special types will now be discussed as a sampling of antenna design variations.

Very-Low-Frequency Antennas

Antennas for use in the low and very-low frequency range (10–300 kHz) are quite complex because of the very long wavelength of signals in this range. A good example is the United States Navy installation at Cutler, Maine. This transmitter, which covers a range of frequencies from 14 to 30 kHz, is used for long-range communications with Navy ships and submarines. The Navy chose this frequency range because of the advantages of vlf propagation.

The antenna arrays for this system resemble a gigantic spider web spun in the sky. The antenna can be visualized as a huge capacitor; the web of the antenna forms one plate of the capacitor, and a screen of wires placed in the earth to reduce losses forms the other. The ground wires extend into the ocean to form a better conductor. Both webs of the antenna cover two square miles.

Long-Wire Antennas

If the antenna is a simple resonant dipole, its radiation pattern has the shape of a figure eight, as discussed earlier. However, if the antenna is made more than a wavelength long, there are four major lobes as well as a number of minor lobes. Several long wires may be connected together so that their pattern lobes add to produce an array with greater gain. There are several ways in which this can be done. Two methods discussed here are illustrated by the "vee" and the rhombic antennas.

The "Vee" Antenna—The basic "vee" antenna is illustrated in Fig. 3-24; each of its two legs measures at least a wavelength. The individual patterns of each leg add, as shown, to produce a front-and-back lobe arrangement that makes this antenna bidirectional. In addition

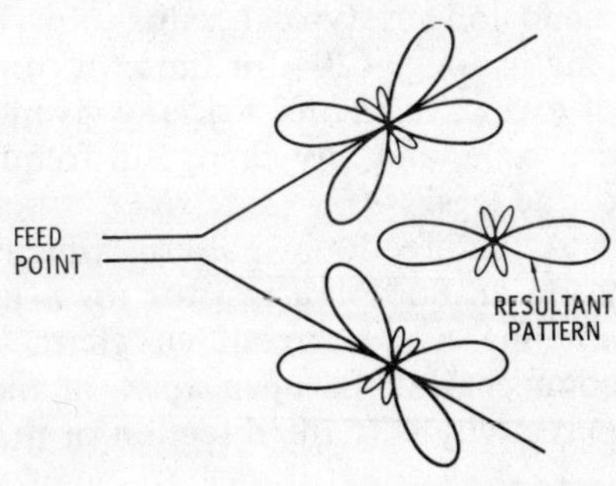

Fig. 3-24. The basic "vee" antenna.

to the two primary lobes, there are important secondary lobes; thus, there can be reception from several directions. This makes orientation of the antenna a problem unless the existence of the secondary lobes is recognized and taken into account. The "vee" antenna has a very broad frequency response but rather poor mechanical characteristics.

The Rhombic Antenna—The rhombic antenna, illustrated in Fig. 3-25, has been used only to a small degree in the vhf band because of its large size requirements. It has characteristics, however, that make it a desirable antenna, and in the uhf band it becomes practical

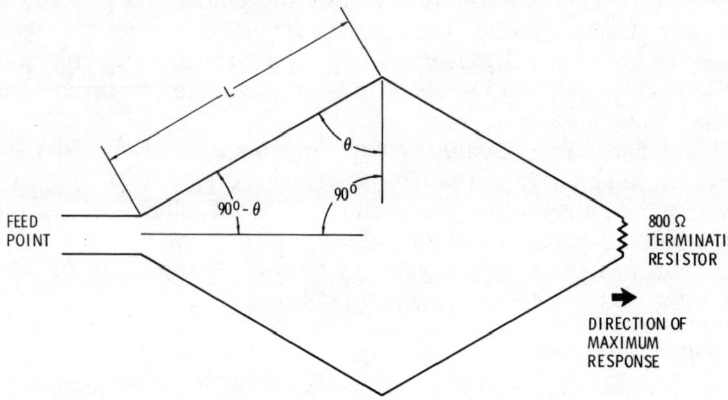

Fig. 3-25. The basic rhombic antenna.

because of the much smaller size requirements. It is a broadband unidirectional unit with a high gain. Each leg (L) in Fig. 3-25 should be several wavelengths at the operating frequency. If the legs are made longer, the rhombic will have even greater gain and sharper directional characteristics. Because of this, the unit has a gain curve that increases with frequency, since as the wavelength of the signal becomes shorter, the legs become electrically longer.

The tilt angle (θ) varies with the length of the legs in order to maintain a proper balance. This angle becomes greater as the legs are made longer; typical values are 130° for legs that are four wavelengths long, 120° for those of three wavelengths, and 100° for those of two wavelengths. For any given installation, the tilt angle (θ) stays the same, and any change in frequency effectively changes the length of the legs.

A unidirectional pattern, which is usually desired, is obtained by using an 800-ohm resistor for a termination; without it the rhombic antenna has bidirectional characteristics. The 800-ohm resistor is connected at the open apex, at the opposite end from the feed line. Directivity is in the direction of the resistor.

If a terminating resistor is added, the rhombic antenna not only becomes unidirectional, but also has an approximately uniform current distribution along the legs. It does not resonate at any one frequency and thus responds well over a wide frequency range. In the uhf television band, where the range from highest to lowest frequency is less than 2 to 1, a single well-designed terminated rhombic can be used for complete coverage. If each leg is made slightly longer than 2 feet, the antenna will be about two wavelengths at the high end of the band and about one wavelength at the low end. Longer lengths will, of course, mean greater gain at both ends.

Stacking can be used to increase the gain, as with other antennas. Normal separation is about one-half wavelength (at the center frequency of operation) between sections. Each stack increases the gain and makes a sharper forward lobe. Stacking also increases the vertical directivity and helps eliminate interference from ground reflections.

The Yagi Antenna

In general, any antenna array having one active dipole element and two or more parasitic elements (one reflector and one or more directors) is known as a *Yagi* antenna after the Japanese physicist, Hidetsugu Yagi, its inventor.

The basic Yagi is one of the highest-gain antennas. However, because of its resonant nature, its frequency coverage is somewhat restricted. Several factors affect the frequency response of any antenna. Among these are the number of elements, their diameter, and the spacing between them. Thus, even the Yagi, although basically a resonant antenna, can be constructed to have a wider or narrower frequency coverage depending on the type of service for which it is intended.

As more elements are added to a Yagi antenna, the impedance of the driven element is lowered. This is true of any Yagi system; signal addition from the nondriven elements increases the signal currents in the driven element, thus greatly reducing the impedance.

Typical impedance values are from 4 to 16 ohms for the center of the driven element of a 6- to 10-element Yagi. For a simple dipole, the center-fed impedance is 72 ohms; as the feed points are moved farther apart toward the ends, the impedance becomes higher. This may be seen from the voltage and current curves of a typical dipole, shown in Fig. 3-26. There is maximum current and minimum voltage at the center feed points, as shown at A. As we move out toward the ends of the elements, the voltage increases while the current decreases; and, since voltage divided by current equals impedance, the impedance increases as the feed points are moved toward the ends. Thus, the impedance at B is greater than the impedance at A, and at C the impedance is greater than it is at B.

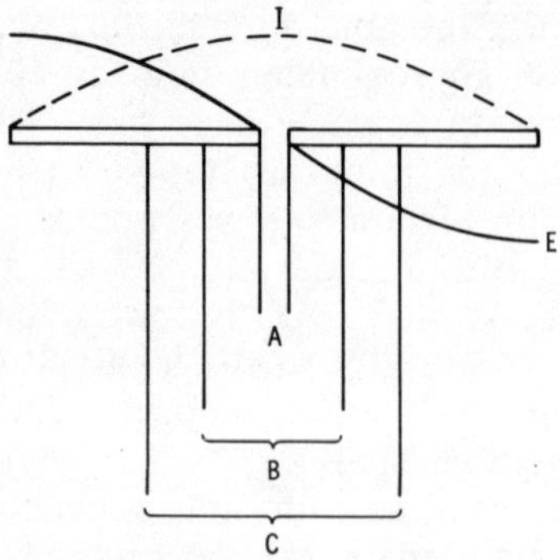

Fig. 3-26. Impedance of dipole increases as feed points are moved farther apart.

The low-center impedance of a Yagi can be increased by a "Y" or "delta" feed as in Fig. 3-27, which matches the typical 300-ohm transmission line to 300-ohm impedance points at the dipole element. The use of a folded dipole with a segment of open-wire transmission line is also a common means of matching the Yagi to the common 300-ohm line.

The pattern of a typical Yagi has a sharp forward main lobe, with much smaller secondary and rear lobes. Because of this sharp pattern, the Yagi requires careful orientation. For best results, the station must be in a direct line with the direction of greatest pickup. Typical Yagi antennas are shown in Fig. 3-28.

Log-Periodic Antennas

A recently developed antenna used for fringe-area television reception is the log-periodic antenna. This design is based first of all on the "transposed-harness" concept. This concept incorporates a large number of active dipoles connected together in a special "cross-phasing" arrangement (Fig. 3-29A) which provides a progressive 180° phase shift between elements. A "split-boom" or "double-boom" arrangement, shown in Fig. 3-29B, accomplishes the same 180° phase shift as the transposed harness. The dipole lengths and their spacings are repeatedly changed at the same constant ratio in accordance with a mathematical formula based on the theory of an infinite spiral. This provides an antenna that has a very wide frequency response and excellent directional characteristics.

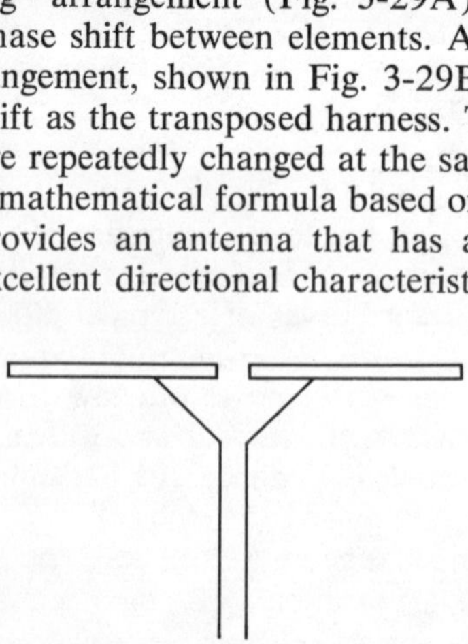

Fig. 3-27. "Delta" or "Y" feed method for impedance matching of transmission line to dipole.

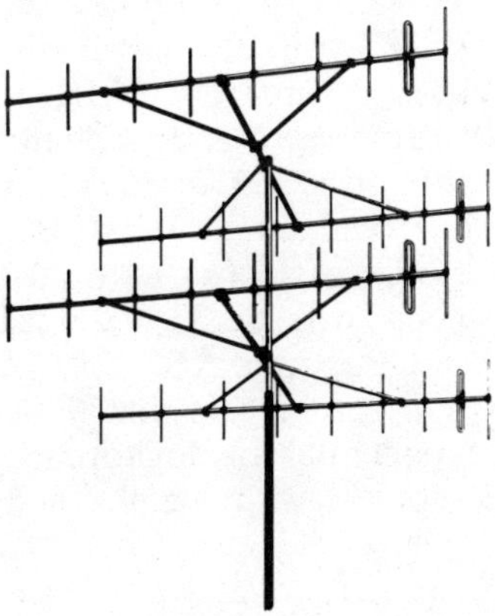

TYPICAL FOUR STACK YAGI ARRAY
WITH NINE ELEMENTS

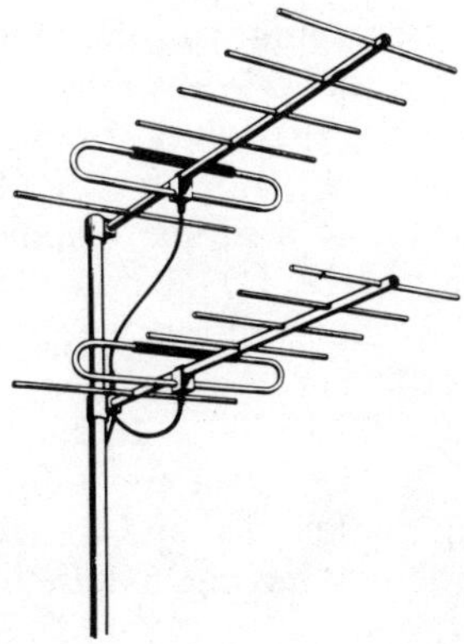

TYPICAL TWO STACK YAGI ARRAY
WITH SIX ELEMENTS

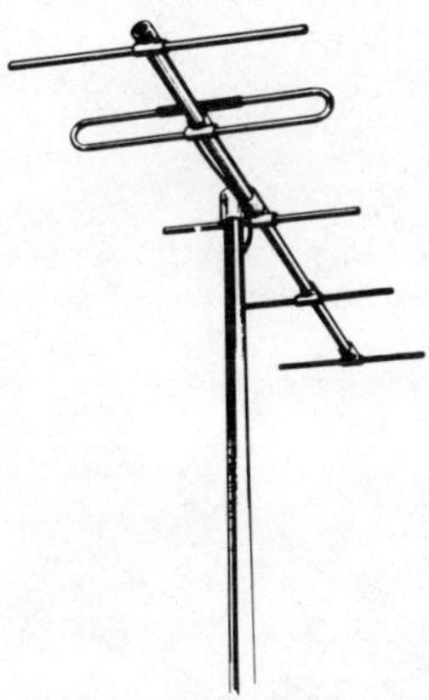

TYPICAL SINGLE STACK YAGI
WITH FIVE ELEMENTS

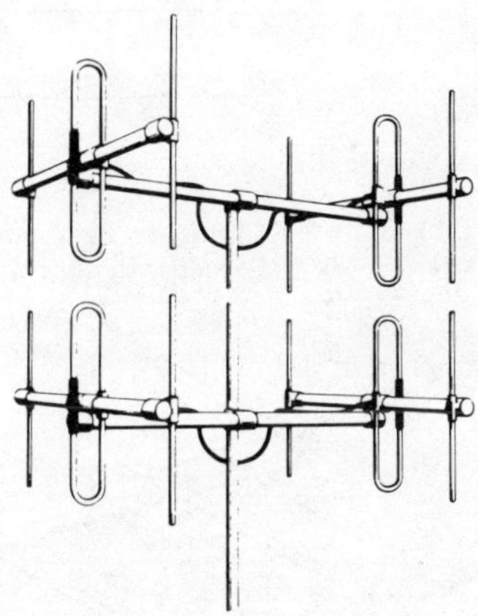

TYPICAL FOUR STACK YAGI ARRAY
WITH THREE ELEMENTS

Court sy Sinclair Radio Labs, Inc.

Fig. 3-28. Typical Yagi antennas.

Beginning with the assumption that the dipole length ratio, L5/L6, will equal 0.9, the inverse frequency ratio, F6/F5, will also equal 0.9. If F6 is the lowest frequency of the uhf television band (470 MHz), L6 is the length of a dipole resonating at that frequency and L5 is the length of a dipole resonating at 522 MHz. In turn, L4 resonates at 580 MHz, L3 at 644 MHz, L2 at 716 MHz, and L1 at 795 MHz. Since these dipole lengths are determined by the angle (a) they subtend from the array apex point, the spacings will be S1 = 0.9S2, S2 = 0.9S3, etc. It can also be shown that the ratio (S2 − S1)/(L2 − L1) is a constant, k. Mathematically it can be said that these dipole lengths and spacings follow an exponential or logarithmic function having a base of k. Additional dipoles, with lengths and

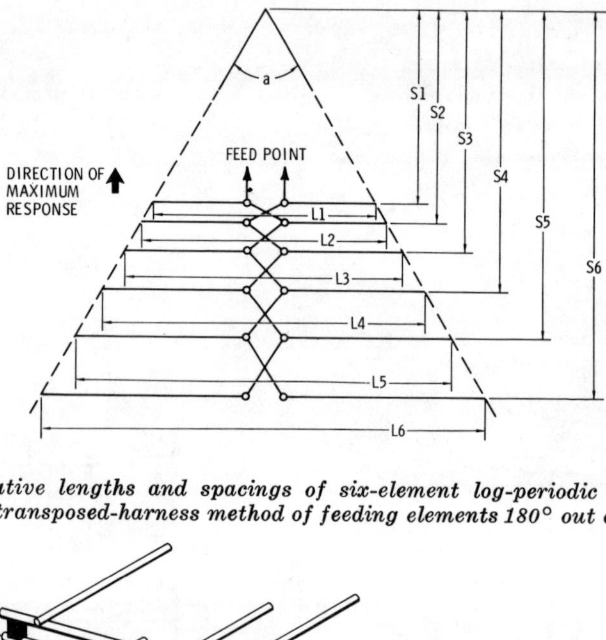

***(A)** Relative lengths and spacings of six-element log-periodic antenna showing transposed-harness method of feeding elements 180° out of phase.*

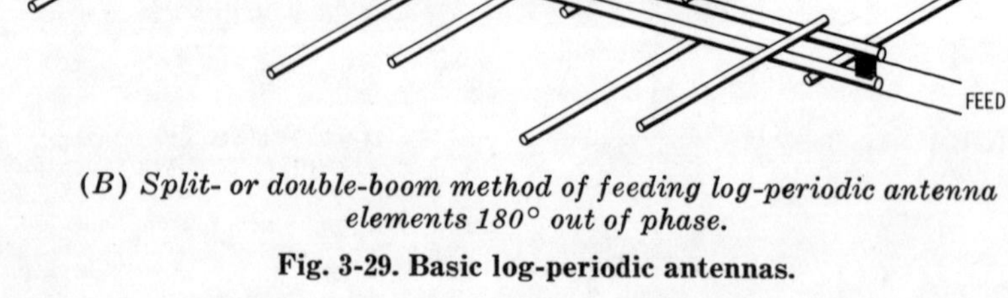

***(B)** Split- or double-boom method of feeding log-periodic antenna elements 180° out of phase.*

Fig. 3-29. Basic log-periodic antennas.

spacings also following the logarithmic function having a base of *k*, give the log-periodic antenna the capability of also covering the vhf television channels 2 through 13. Fig. 3-30 shows a commercial log-periodic antenna capable of receiving all tv channels from 2 through 83.

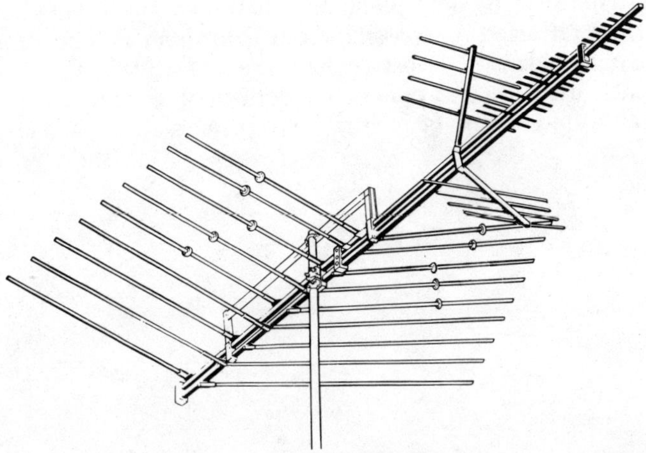

Courtesy JFD Electronics Components Corp.

Fig. 3-30. Log-periodic antenna used for reception of tv channels 2 through 83.

ANTENNAS FOR MICROWAVES

Microwave antennas consist in general of a transmission or feed system and a focusing device. The transmission system is usually a hollow, rectangular pipe or *waveguide* rather than coaxial cable or two-wire open transmission line. The reason is that at microwave frequencies, much higher losses are experienced with coaxial cable and two-wire line than with waveguides.

The focusing device is mounted in such a manner that it collects radiation from the open end of the waveguide and focuses it into a relatively narrow beam of energy for transmission. For receiving a signal, the focusing device collects incoming radiation from a transmitting antenna and focuses it into the waveguide. The focusing device is usually parabolic in shape; the reason for this has been discussed previously.

Microwaves are usually considered as being in that band of radio frequencies above 1000 MHz. This covers part of the ultrahigh frequency band (uhf), which is from 300 to 3000 MHz; all of the superhigh frequency band (shf), which is from 3000 to 30,000 MHz; and all of the extremely high frequency band (ehf), which is from 30,000 to 300,000 MHz. At a frequency of 3000 MHz, the wavelength is 10 centimeters (about 4 inches). Circuit and antenna elements of about this size and smaller are common in microwave equipment.

In many ways, microwaves behave similar to visible light waves; one characteristic is straight-line, or "line-of-sight," propagation. Both light waves and microwaves tend to travel in straight lines and are easily reflected by both plane and curved surfaces. They also can be bent, or refracted, by passing from a medium of one density to that of another density. Lenses, which use refractive optics, and mirrors, which use reflective optics, are both used as antennas to focus microwaves into beams of energy for transmission from one station to another.

4

Base-Station Antennas

The base station is the primary link of most communications systems. While there may be radio communication between mobile units, it is usually the base station that is of greatest importance for the complete system. In turn, the base station must have an antenna providing ample and proper radiation. All base-station antennas have some similar characteristics. However, the nature of amateur, Citizens band (CB), and business radio also dictates many differences. Amateur radio, although a very important factor in experimental work and in times of disaster, is basically a hobby of those involved, while business radio is often a key portion of earning an income. Citizens band may be used for personal and business reasons but is not intended primarily for either.

AMATEUR ANTENNAS

Economy, time, and space are factors often considered in the choice of an amateur base-station antenna. Since amateur radio is a hobby, the amateur usually spends money for equipment only after the necessities of life have been provided for. This may mean using a length of stranded copper wire instead of the more desirable plated fiber-glass rod.

Most amateurs spend hours building and rebuilding their equipment, and nearly every contact made with another amateur, directly or via their radio waves, and the monthly *QST* magazine (official amateur publication) bring news of equipment or antenna variations. Another challenge to the amateur operator is that of space. In years past, the amateur nearly always had a lawn, a house roof, and possibly the neighbor's roof, for antenna construction. Today, his space may be limited to that of a windowsill in an apartment complex. While the windowsill doesn't provide for much leeway, the almost endless variation in antennas used by amateurs does. The following paragraphs will discuss the basic types of antennas used with amateur base stations.

Horizontal Antennas

The most common and the most practical antenna for the 3.5-MHz (80-meter) and 7-MHz (40-meter) amateur bands is the half-wave horizontal dipole. Many forms of dipoles can be constructed, and many variations are possible for coupling the transmission line to the antenna. Sometimes it is not possible to feed the antenna through a transmission line—for example, when there is not enough space to erect a horizontal dipole with a transmission line, or when operation on a higher harmonic is desired. The use of a single wire without a transmission line is no longer considered good practice because it can cause severe interference to television receivers (tvi). It can also cause broadcast interference (bci) to neighboring radio receivers.

Fig. 4-1 shows two end-fed *Hertz* antennas. One end of the antenna wire connects directly to the tvi filter, meaning that a portion of the antenna is inside the transmitter room. Ideally the full antenna wire is kept straight and horizontal. However, either end of the wire can be bent and raised diagonally as shown to fit the space available.

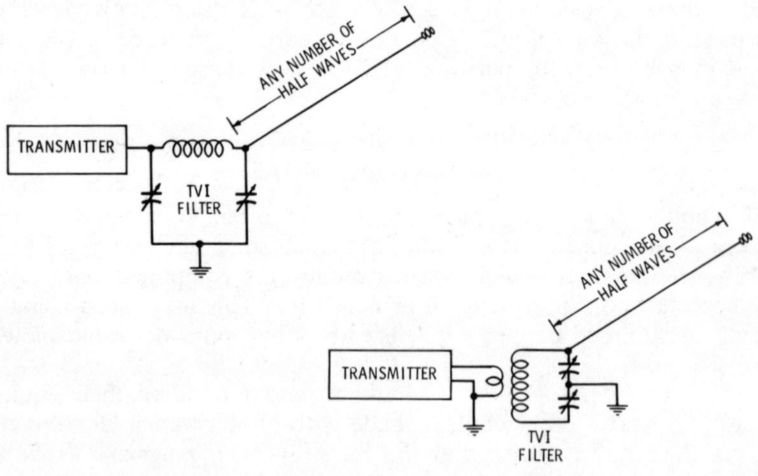

Fig. 4-1. Two methods of coupling an end-fed Hertz antenna to a transmitter.

The length of the antenna wire can be any multiple of a half wave. The length of a half wave in space is based on the simple formula:

$$\text{Length (in feet)} = \frac{492}{\text{Freq (MHz)}}$$

However, because wire thickness, capacitive effects, and other factors must be considered, the following formula is used for determining the physical length of a half-wave antenna:

$$\text{Length (in feet)} = \frac{492 \times 0.95}{\text{Freq (MHz)}} = \frac{468}{\text{Freq (MHz)}}$$

Thus, a half-wave antenna for 14.2 MHz would be $\frac{468}{14.2} = 33.6$ feet. When installing an end-fed Hertz antenna, the wire should enter the building through ceramic or glass insulators. At least three-quarters of the wire should be outside the building and as high off the ground as possible. This antenna can be operated on its higher harmonics, or it can be used as a quarter-wave Marconi antenna at half its frequency when used with a ground system. Two methods for coupling the antenna to a transmitter are shown in Fig. 4-1. A tvi filter should be inserted between the transmitter and the coupling network.

Where space is limited or when multiband operation is desired, the *Zepp* antenna (Fig. 4-2) has worthwhile advantages. It can be used

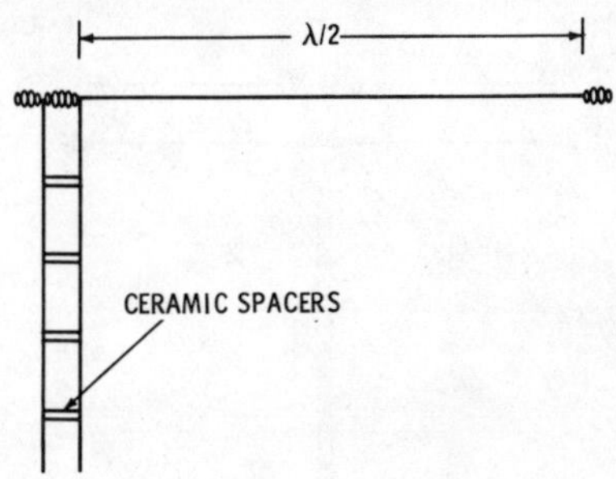

Fig. 4-2. Voltage-fed *Zepp* antenna.

on several bands by simply retuning its feeders. The radiating portion of this antenna is always some multiple of a half wavelength long. A high voltage is present where the radiating portion of the antenna connects to the feeders. Accordingly, it is called a *voltage-fed* antenna.

A center-fed antenna displays improved radiating capabilities because the feeder (transmission line) is balanced to ground electrically. This means that the feeders will not radiate excessively when the system is properly tuned. Thus, the antenna will radiate all or most of the energy. There are several methods of feeding the antenna at its center. Fig. 4-3 shows one of them—the *center-fed Zepp*.

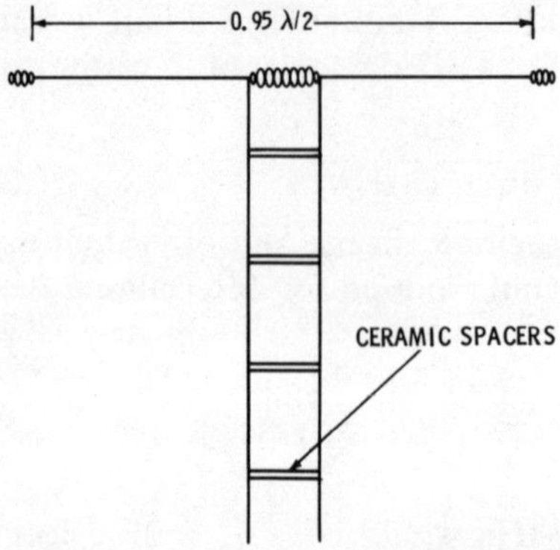

Fig. 4-3. Center-fed *Zepp* antenna.

The center-fed Zepp, commonly known as the *current-fed doublet* antenna, is called a balanced system because the wires on each side of center are of equal electrical length. This antenna is very popular because it can operate over a wide range of frequencies when both the feeders and the antenna wires can be tuned to resonate at a particular frequency. This antenna, when a half wave long (radiating wire) or when used on its odd harmonics, is known as a current-fed system. If tuned for operation on its even harmonics, it is known as a voltage-fed system.

A half-wave doublet can also be fed with coaxial cable (Fig. 4-4). Either 52-ohm or 75-ohm cable may be used; usually, the 75-ohm

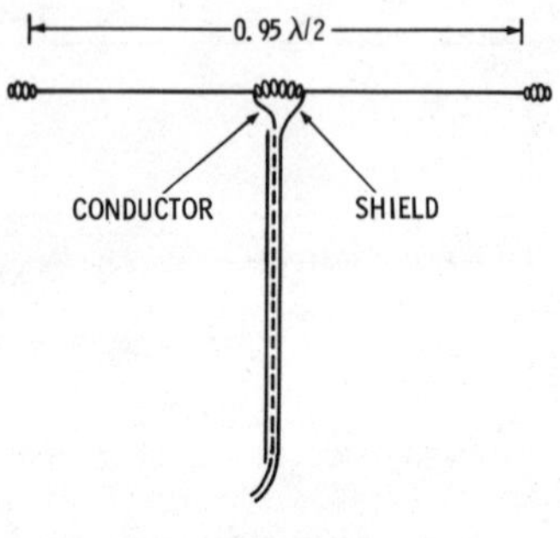

Fig. 4-4. Half-wave doublet antenna fed with coaxial cable.

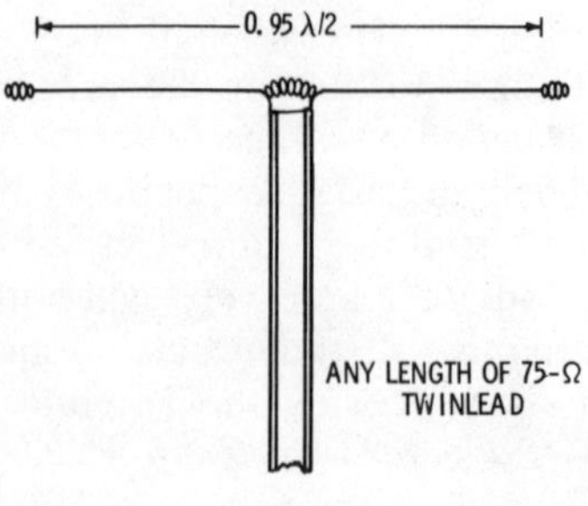

Fig. 4-5. Half-wave doublet antenna fed with 75-ohm twin lead.

type is preferred because it matches the impedance of the system better than the 52-ohm type. Because of the asymmetry of this system, unwanted radiation from the coax shielding is likely to occur. The use of twin lead is a better choice.

Special polyethylene-dielectric twin lead has been developed and is very popular for feeding doublet-type antennas because of the very low losses of the material. The standard doublet antenna shown in Fig. 4-5 uses 75-ohm twin lead for its transmission line. The length of the line is not critical and can be matched to almost any tuning system or circuit.

Courtesy Cush Craft Corp.

Fig. 4-6. Stacked, horizontal, 3-element Yagi array for the 144-, 220-, and 420-MHz amateur bands.

Several other methods are used for feeding the half-wave dipole. These require the application of special formulas for each type and are covered in detail in the amateur radio handbooks.

There are many other types of horizontal antennas used by the amateur operator. Among these are the folded dipole, various fed and parastic arrays (including the Yagi in its many forms), and stacked arrays. A good many of these depend upon the frequency band being used. That is, the Yagi that would be mechanically impossible because of size at 4 MHz is very practical at 28 and 50 MHz. And at 144, 220, and 420 MHz, the Yagi and the stacked Yagi have limitless possibilities. Fig. 4-6 shows a stacked, horizontal, 3-element

Yagi array for the 144-, 220-, and 420-MHz amateur bands. Collinear arrays, log-periodic antennas, and antennas with parabolic reflectors are also possible in the higher amateur frequencies.

Vertical Antennas

The classic example of the vertical antenna is the half-wave antenna tower used as the radiator of signals in the a-m broadcast band. The half-wave vertical antenna is the most effective for the relatively short-range ground-wave propagation employed at broadcast frequencies and also in the 1.8-MHz (160-meter) amateur band. To be effective, the half-wave vertical must be used where ground conductivity is good. Two methods for feeding the vertical antenna are shown. Fig. 4-7A shows how a 75- to 600-ohm line is coupled to the

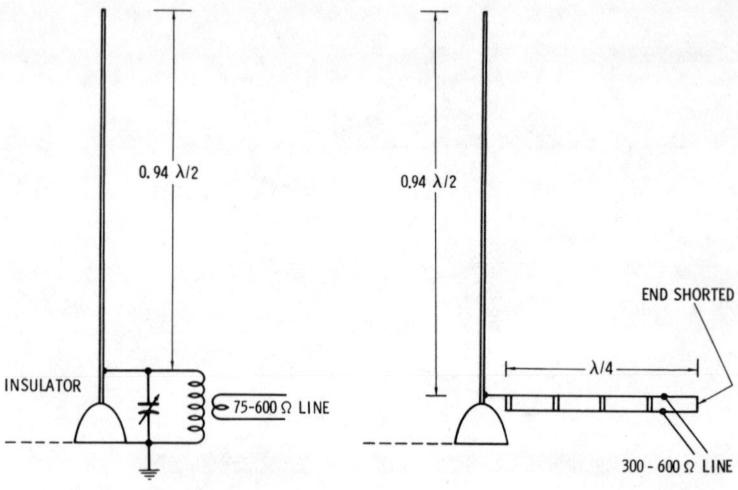

(*A*) *Transformer coupling.* (*B*) *Quarter-wave matching stub coupling.*

Fig. 4-7. Two methods of feeding a half-wave vertical antenna.

vertical radiator. The tuned circuit is located at the base of the antenna. Note the ground (earth) connection. The second method (Fig. 4-7B) is called a stub-fed vertical, where the half-wave radiator is fed by a quarter-wave stub and a 300- to 600-ohm transmission line.

The Marconi (discussed earlier and illustrated again in Fig. 4-8) is a grounded quarter-wave vertical antenna. The radiating portion is above ground, and the "image" antenna is represented by the effect of the ground below the quarter-wave radiator section. The action of this antenna is similar to that of a half-wave antenna. The radiating pattern is one-half the vertical plane pattern of a free-space dipole.

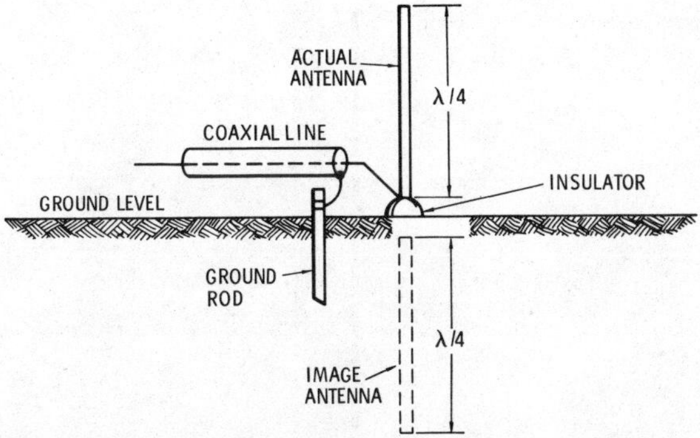

Fig. 4-8. The Marconi, or grounded quarter-wave vertical antenna.

A Hertz antenna is basically a half wavelength long, while the Marconi is one-quarter wavelength with one end grounded.

The electrical length of the Marconi antenna can be increased or decreased by using series-loading coils or a capacitor, as shown in Fig. 4-9. This technique makes it possible to improve the radiation pattern of the antenna.

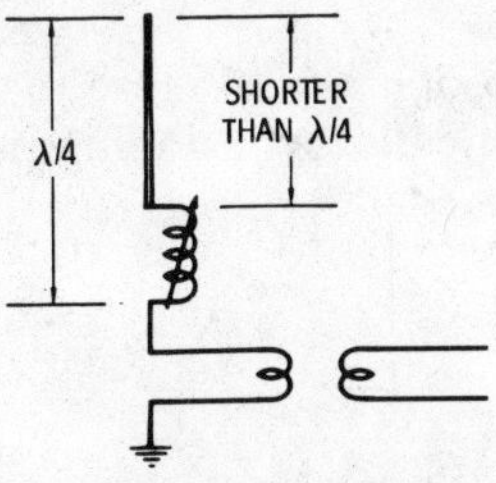

(*A*) *Inductive loading increases the electrical length.*

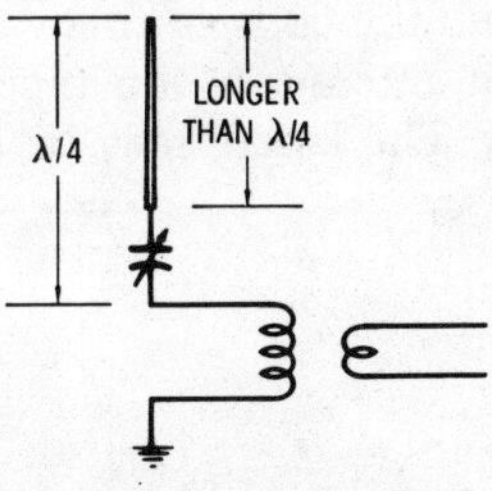

(*B*) *Capacitive loading decreases the electrical length.*

Fig. 4-9. Two methods of changing the electrical length of a Marconi antenna.

Most of the vertical antennas designed for use on the amateur bands are of the Marconi resonant quarter-wave type. This type of antenna must work against a good ground system to furnish the other quarter wave to complete the dipole antenna. A series of radial wires, called a *counterpoise,* may be used where the ground lacks good conductivity. As shown in Fig. 4-10, when the antenna is made one-quarter wavelength, the impedance between the base of the antenna and ground will be from 40 to 50 ohms. By connecting 52-ohm coaxial transmission line at this point, a good match of line to antenna is achieved over a considerable portion of the band for which the an-

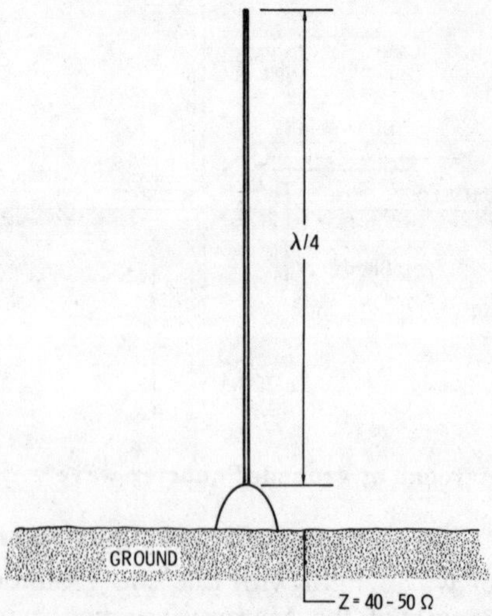

Fig. 4-10. The impedance between the base of a Marconi antenna and ground is approximately 40 to 50 ohms.

tenna is made one-quarter wavelength. This match is done without any tuning coils or other matching devices.

An antenna of this type can be made to operate on more than one band. The length may be adjusted automatically by installing parallel

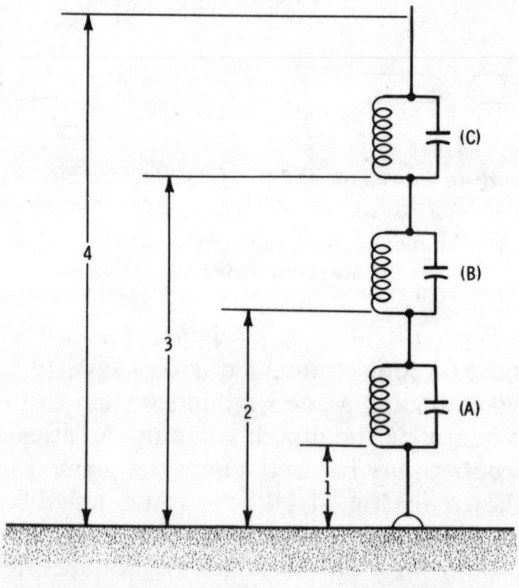

Fig. 4-11. Installing parallel resonant traps enables the Marconi antenna to operate on more than one band.

resonant traps at the proper points to cut the antenna electrically at one-quarter wavelength, as in Fig. 4-11. These traps offer a very high impedance at or near resonance and act as insulators placed at the end of the quarter-wave point for each band. In Fig. 4-11, for example, length 4 includes the four antenna sections and the inductances of traps (A), (B), and (C) to give an electrical length equivalent to one-quarter wavelength at 28 MHz. Resonating at 50 MHz, trap (C) reduces the effective height to one-quarter wavelength at that frequency. In turn, traps (B) and (A) reduce the effective height to that of one-quarter wavelength at 144 and 220 MHz, respectively.

Other examples of vertical antennas used for amateur base stations include the half-wave vertical with tuning/matching ring shown in Fig. 4-12; the stacked, vertical, 11-element Yagi array for the 144-MHz band shown in Fig. 4-13; and the vertically mounted half-wave dipoles for the 144-, 220-, and 420-MHz bands shown in Fig. 4-14.

Ground Systems

As mentioned previously, vertical antennas like those used in the a-m broadcast and lower-frequency amateur bands rely heavily upon

Courtesy Cush Craft Corp.

Fig. 4-12. Half-wavelength antenna with tuning/matching ring.

Courtesy Cush Craft Corp.

Fig. 4-13. Stacked, vertical, 11-element Yagi array for the 144-MHz amateur band.

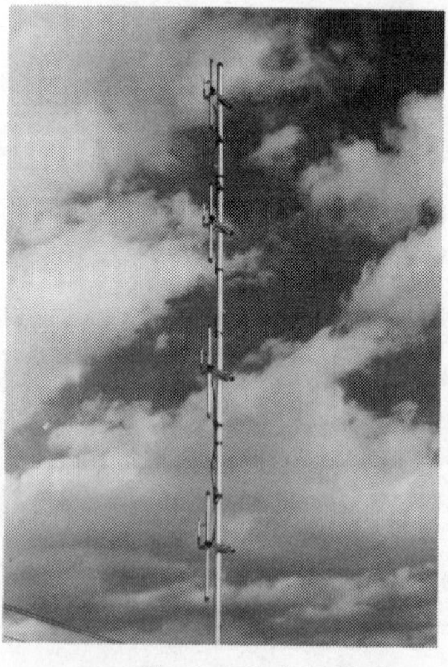

Fig. 4-14. Vertically mounted half-wave dipoles for the 144-, 220-, and 420-MHz amateur bands.

Courtesy Cush Craft Corp.

a good ground system for maximum coverage. The reason is that ground-wave propagation is the primary means of communicating at these frequencies. In order to have a good ground system, the antenna first of all should be erected where ground conductivity is good. Low ground, where the moisture content is higher, provides a more ideal location for this type of antenna than high ground, where the conductivity is poorer because of the drier earth.

Regardless of where the antenna is located, it is almost always desirable and thus customary to provide an extensive system of wires buried in the ground and extending radially outward from the base of the antenna. This reduces ground losses in the immediate vicinity of the antenna, where current density is highest.

Ground systems recommended by the Federal Communications Commission (FCC) for broadcast stations consist of a minimum of 120 radials, each as long as the antenna height. These extend from the base of the perimeter of a complete circle around the antenna. The higher frequencies used by the amateur permit a minimum of four radials about as long as the equivalent length of the antenna (Fig. 4-15). Additional radials will improve the performance. Copper rods should be secured to the ends of each radial and driven into the ground. A ground rod should also be connected to the antenna base as shown. If there is insuffícent space to install radials of the length recommended, they may be bent back slightly, or cut shorter and more radials added to the system. If the antenna site is rectangular rather than square, the ground system will still be satisfactory as long as it covers about the same area and is installed in the recommended manner.

All antenna installations must be protected again lightning. A properly designed and installed horn or ball-gap device at the base (Fig. 4-16) will protect the transmitter building and equipment and will meet Underwriters requirements. A gap of ⅛ inch between wires or balls will not flash over at the maximum power permitted the amateur operator.

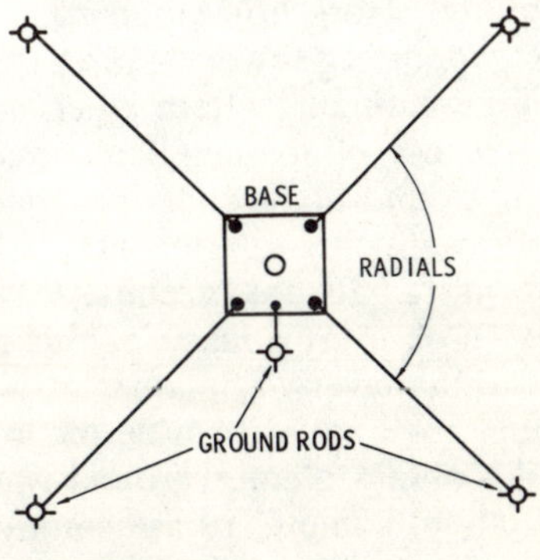

Fig. 4-15. Radially positioned ground wires reduce ground losses.

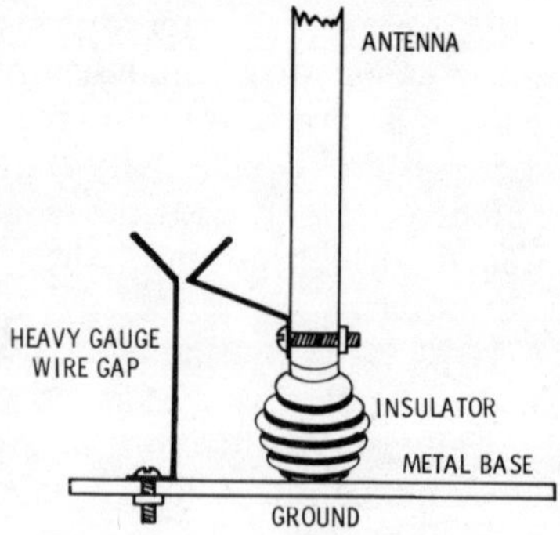

Fig. 4-16. Method of protecting vertical antenna installations against lightning.

CITIZENS BAND ANTENNAS

Operating within the 23 channels between 26.965 and 27.255 MHz, Class D CB base-station antennas are similar to those used on the 28- to 29.7-MHz (10-meter) amateur band. Either the half-wavelength dipole or the quarter-wavelength vertical antenna will cover the entire range adequately. While most CB base stations use vertical antennas, there is some usage of horizontal and combination vertical/horizontal antennas. Use of the horizontal antenna provides some freedom from excessively used channels by not responding to vertically polarized waves. However, the horizontal base-station antenna does not provide good communication with mobile units using vertical antennas. To obtain extended coverage with the maximum allowable power, driven and parasitic arrays are used.

Vertical Antennas

A basic ground-plane antenna is shown in Fig. 4-17. The radiating element is a quarter-wave vertical. Below the radiator are three or four quarter-wave rods (radials). These project downward at an angle slightly below horizontal. The angle is selected to maintain the correct impedance match to the system. When additional isolation is required, more horizontal rods can be mounted at a lower point on the supporting tower or pipe. The ground-plane antenna produces an omnidirectional pattern when mounted as shown.

Because of its simplicity and economy, the quarter-wave vertical with ground-plane radials is the most common type used for CB base stations. Such a quarter-wavelength antenna has a height of 8.67 feet, and its radials are of equal length. A 50-ohm coaxial cable matches the 36-ohm impedance of this antenna to the 75-ohm impedance of the transceiver. The quarter-wave vertical provides an omnidirectional coverage with little or no effective gain.

To obtain greater ground-wave coverage while reducing sky-wave radiation, some CB base stations use half-wavelength vertical anten-

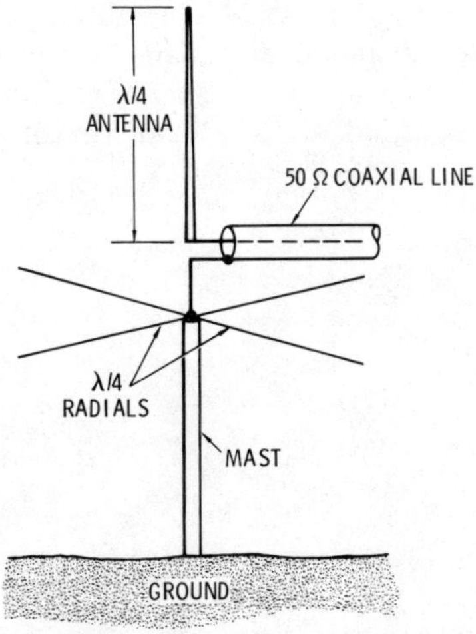

(A) Schematic representation.

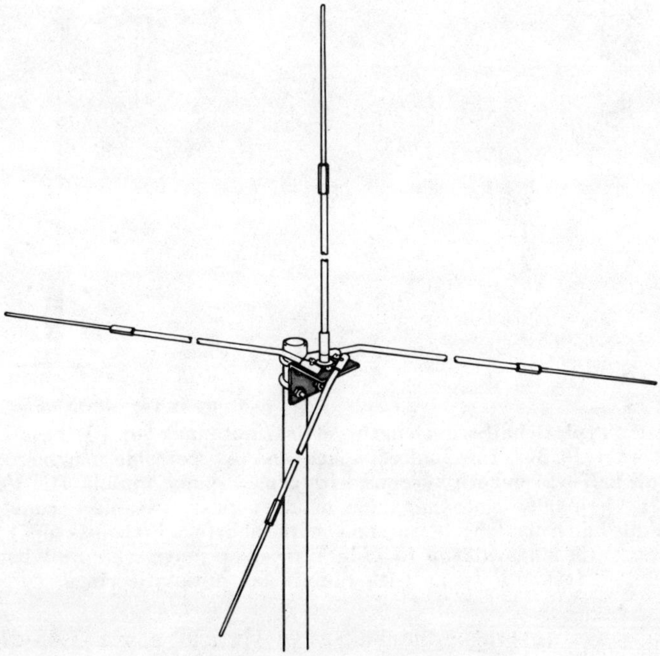

Courtesy The Antenna Specialists Co.

(B) Physical appearance.

Fig. 4-17. Basic ground-plane antenna.

nas. With heights of better than 17 feet and equal radials serving as a ground plane, this half-wavelength vertical is rather large. Inductance coils are often used in both the vertical radiator and the ground-plane radials to physically reduce their lengths. By extending the

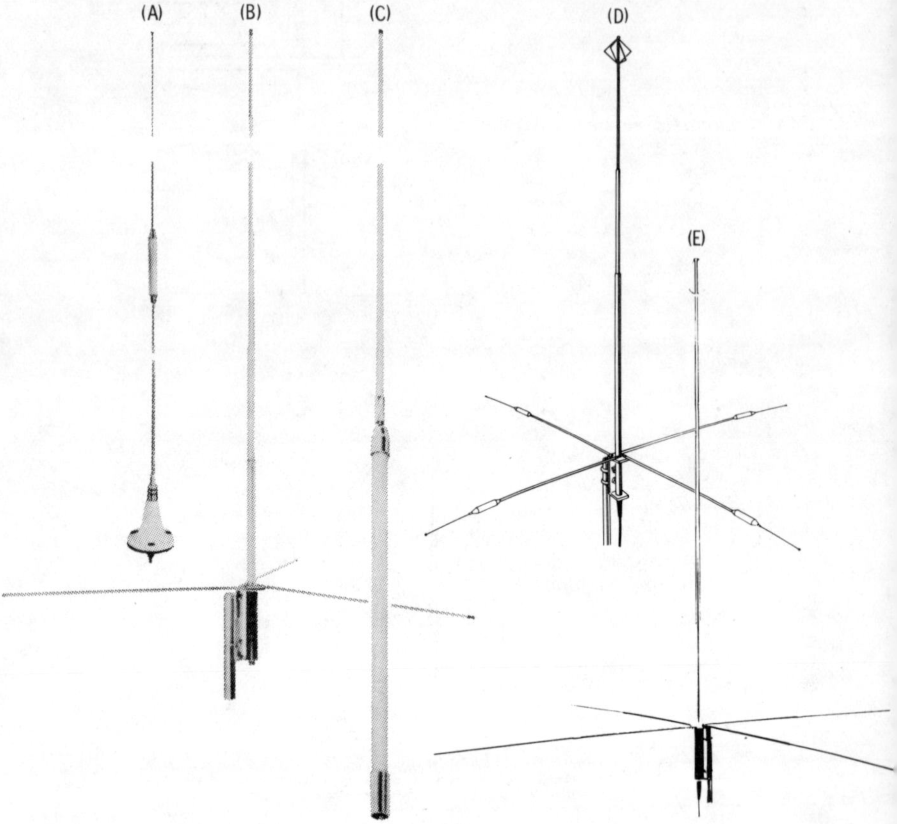

Courtesy The Antenna Specialists Co.

Fig. 4-18. Typical half-wavelength vertical antennas for CB base stations. (A) Center loaded for limited space and as portable marine antenna. (B) Full half-wavelength antenna with ground-plane radials. (C) Full half-wavelength marine antenna made of fiber-glass, two-piece construction. (D) Full half-wavelength antenna with shortened (coil-loaded) radials and antistatic rings at top to reduce receiver noise. (E) Full half-wavelength antenna with radials and antistatic rings.

ground-wave coverage, the half-wave vertical has a 3.75-dB gain. For locations where space is limited or temporary in nature, half-wavelength antennas are available without the ground plane. Fig. 4-18 shows some typical half-wavelength vertical antennas used with CB base stations.

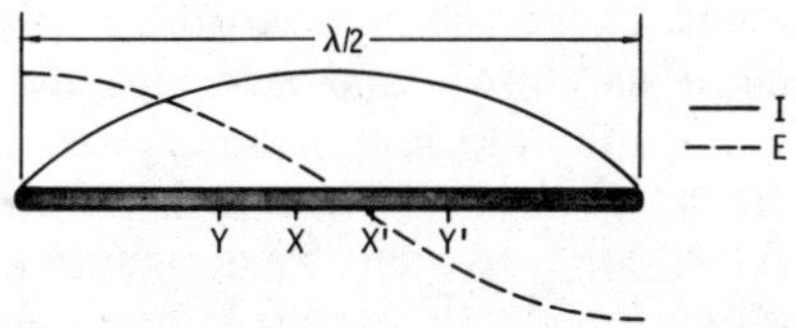

Fig. 4-19. Impedance differences on dipole antenna.

The familiar Yagi antenna with its one driven dipole element and its parasitic elements is the primary directional antenna used for CB base stations. In contrast with the Yagis used for television reception, most of those used for CB are vertical for the obvious reason that vertical antennas of one form or another are used almost exclusively in CB communications with mobile stations. The Yagis built by one manufacturer are listed as having 9.75 to 14.5 dB of forward gain and a 25-dB difference between the forward and the backside radiation.

Because of the effect of the parasitic elements in a Yagi antenna, the effective impedance at the feed point of the driven element is less than that of a simple dipole. Thus, it is necessary to effect an impedance transformation between the transmission line and the driven element. Special matching methods are necessary to do this, since the driven element of Yagi arrays for the CB band (and amateur bands where Yagi arrays are practical) are usually not cut (or separated into two quarter-wave elements) but are single half-wave elements to provide greater mechanical strength. As shown in Fig. 4-19, these impedance-matching methods are based on the fact that as we move outward from the center of the dipole element, the current decreases and the voltage increases as previously explained in Chapter 3. Thus, in Fig. 4-19, the impedance between points Y and Y′ is greater than the impedance between points X and X′.

The most commonly used method of impedance matching when the driven element is fed with 52-ohm coaxial cable is called the *gamma match,* illustrated in Fig. 4-20A. In this system a second ele-

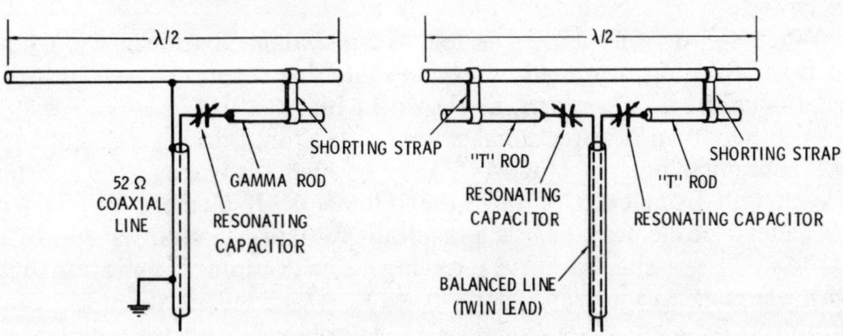

(*A*) *The gamma match.* (*B*) *The "T" match.*

Fig. 4-20. Antenna matching systems.

ment, called the *gamma rod,* is placed parallel with the radiating element as shown and is connected to the radiating element with an adjustable shorting strap. A resonating capacitor is often placed in series with the gamma rod, as shown, so that the inherent inductive reactance presented by the gamma matching section can be cancelled out. The length of the gamma section can also be varied with the shorting strap so that by carefully adjusting both the capacitor and the shorting strap, any reactive component can be eliminated. Thus, the combination becomes a purely resistive load to the transmission line, which reduces the swr on the line to a very low value at the resonant frequency of the driven element. With an antenna cut to its proper resonant length, the gamma match can reduce the swr to less than 1.5:1. A typical five-element vertical Yagi with a gamma match on its driven element is shown in Fig. 4-21.

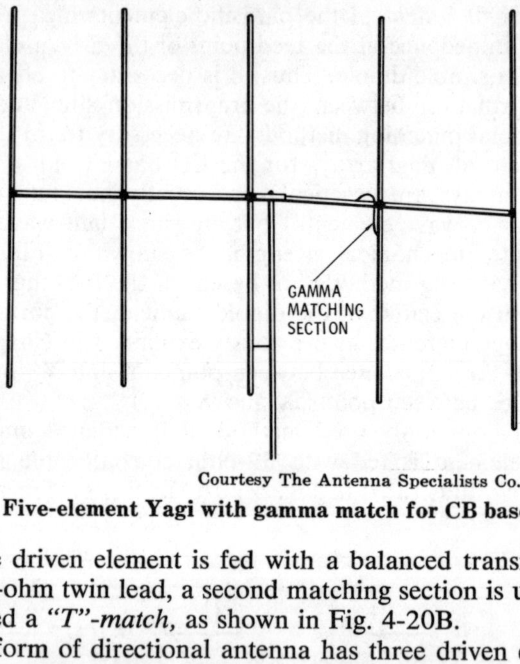

Courtesy The Antenna Specialists Co.

Fig. 4-21. Five-element Yagi with gamma match for CB base station.

When the driven element is fed with a balanced transmission line such as 300-ohm twin lead, a second matching section is used to form what is called a *"T"-match,* as shown in Fig. 4-20B.

Another form of directional antenna has three driven elements set in a triangular pattern (Fig. 4-22). Each element is spaced a quarter wavelength from each of the other elements. If element (1) is fed 90° ahead of element (2), the resultant radiation pattern is that of a cardioid (heart-shaped curve) having its maximum radiation in line with element (1) toward element (2). A second cardioid pattern, produced by elements (2) and (3), is rotated 120° from the first, while elements (3) and (1) produce a third cardioid pattern 120° from the other two. Each cardioid pattern has a beam width of almost

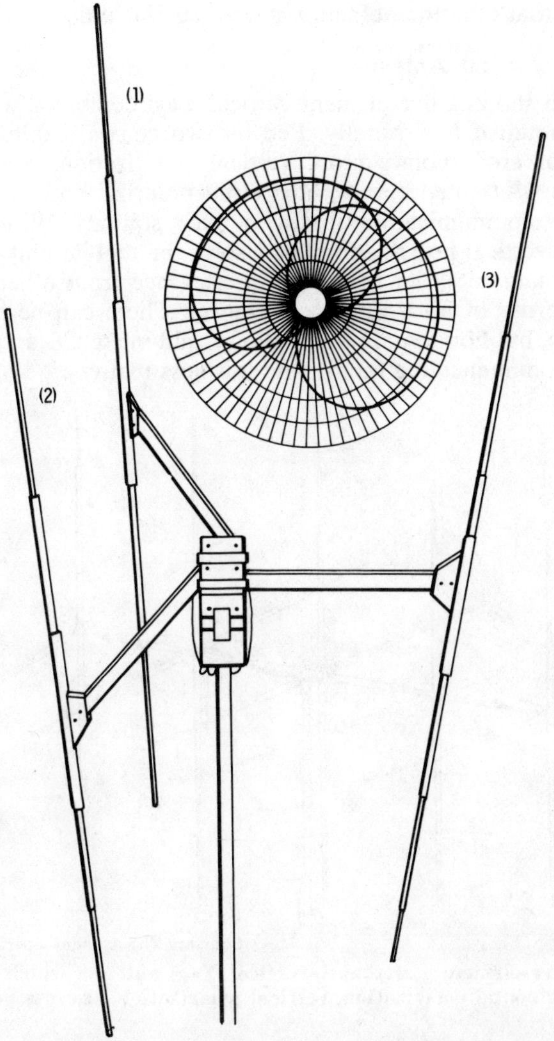

Courtesy The Antenna Specialists Co.

Fig. 4-22. Three-element directional antenna with variable pattern. Note overlapping cardioid patterns shown at top.

180° so that the three cardioids adequately cover the entire 360°. Selection of the cardioid and its direction is accomplished by means of a control box placed near the transceiver. A fourth setting of the control provides an omnidirectional pattern. The cardioid patterns have gains of about 8.75 dB, while the omnidirectional pattern gain is

5.75 dB. This, of course, is the product of one manufacturer, but it is very likely that other manufacturers have similar units.

Horizontal/Vertical Antenna

Fig. 4-23 shows a five-element vertical Yagi combined with a similar Yagi mounted horizontally. Fed by two coaxial cables, a switch provides for cross polarization, vertical polarization, or horizontal polarization. As noted before, horizontal polarization can be advantageous for communications between fixed stations. With so many units using vertical polarization because of the mobile unit space limitations, it may be easier to reduce interference from other CB units through the use of horizontal polarization. There can be no privacy in CB radio, but horizontal polarization could make CB an acceptable means of communicating family and business matters.

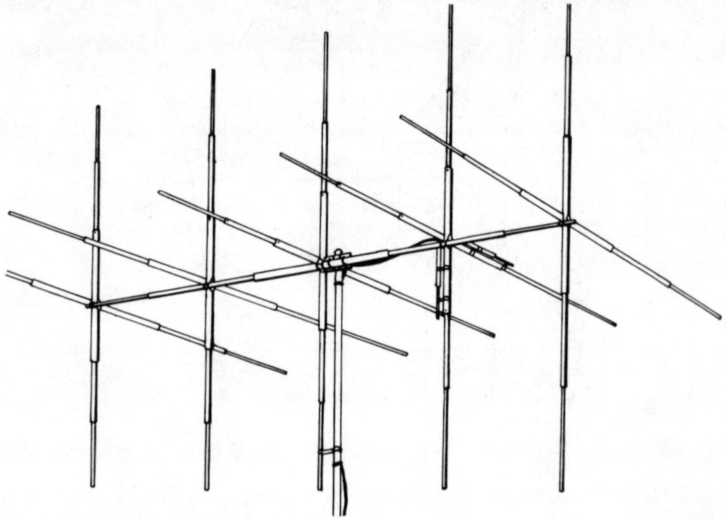

Courtesy The Antenna Specialists Co.

Fig. 4-23. Five-element horizontal/vertical Yagi antenna which provides a choice of horizontal polarization, vertical polarization, or cross polarization.

BUSINESS RADIO ANTENNAS

Reliable communication in business radio systems depends upon the overall effectiveness of both the base station and the mobile unit antennas. The radiation pattern of the base station is extremely important, since signals must be transmitted and received in densely populated areas as well as in the more isolated urban and suburban areas. Many industrial, business, and densely populated areas have intense electrical interference. Such things as motors, arc welders,

X-ray machines, neon signs, fluorescent lights, etc., produce considerable electrostatic and electromagnetic radiation. This unwanted radiation is irregular in form and causes interference on most frequency bands. The radio waves from the business base-station antenna must literally override the interference waves by being 1000 or 10,000 times as large in amplitude. A primary function of the antenna is to concentrate the desired radiation into areas of greatest need. A small plumbing shop, for example, might find it unnecessary to communicate with mobile units in the industrial areas. A large plumbing business doing considerable work within the industrial area, however, must be able to communicate with its trucks working there. Thus, there are many factors to be considered in developing a business radio system.

Determination of the area to be covered gives an indication of the radiation pattern required. If the base station is to be located near the center of the coverage area, a nondirectional pattern, shown in Fig. 4-24, is best. A cardioid pattern will provide best coverage when the base station is at the edge of the area. In other situations, some form of the directional pattern may be desirable.

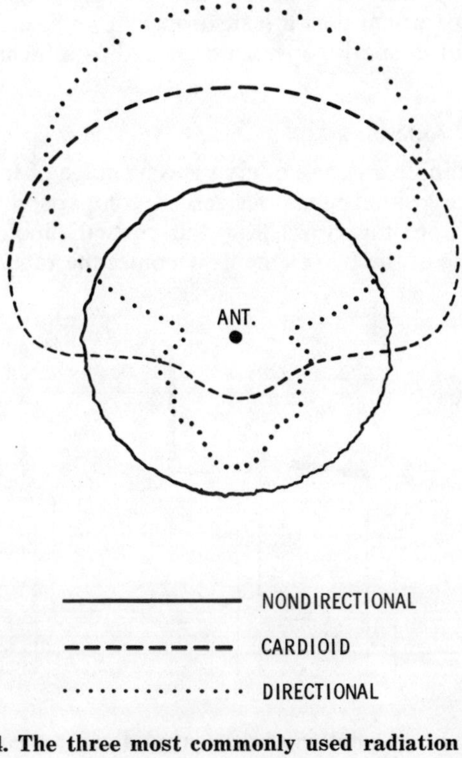

Fig. 4-24. The three most commonly used radiation patterns.

Most business radio operates within two frequency bands—150 to 174 MHz and 406 to 470 MHz. Quarter-wavelength elements for these bands range from about 16 inches to 19 inches and from about 6 inches to 7 inches. Thus, these frequencies, and the corresponding short antenna elements, provide for a great deal of variation in antenna systems. In the 406- to 470-MHz band, the most complex Yagis and collinear arrays become very feasible.

Ground-Plane Antennas for Business Radio

Much like the ground-plane antennas used by CB or amateur systems, the ground-plane antenna for business radio has a quarter- or half-wavelength vertical radiator and three or four ground-plane radials of equal length. Because these antennas are manufactured for the lowest frequency of a given range, charts are provided for cutting to the specific frequency. These antennas for business radio are built for greater durability and higher rf power capacity than similar antennas used in CB radio. The radiators and radials are either cadmium-plated steel, aluminum rod, chrome-plated brass, or stainless steel. Having a capacity of up to 1000 watts, an swr of less than 1.5:1 is obtainable when the 50-ohm impedance is correctly matched (the half-wavelength vertical uses a transformer at its base). All such antennas have built-in spark gaps or dc ground as a means of lightning protection.

Coaxial Dipole Antennas

The coaxial dipole antenna offers the advantage of lower resistance to wind than the ground-plane designs. A comparison with the horizontal doublet antenna shows how the coaxial dipole was evolved (Fig. 4-25). One of the two elements becomes the radiator. The other

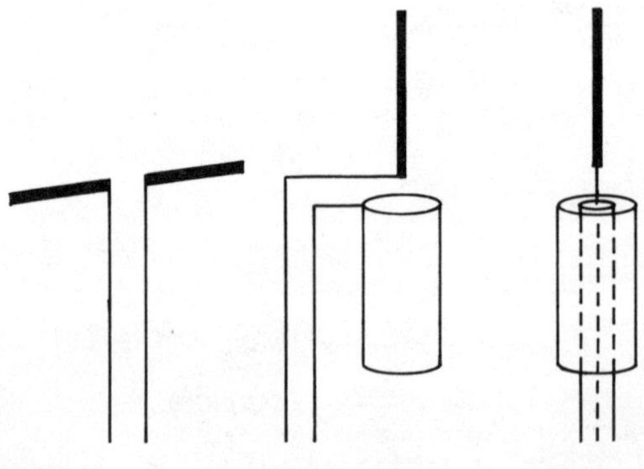

Fig. 4-25. Evolution of the coaxial dipole antenna.

Fig. 4-26. Typical coaxial dipole base-station antenna for the 150–174-MHz business radio band.

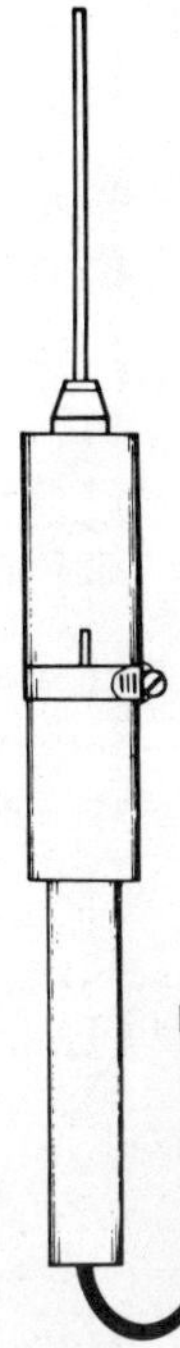

Courtesy The Antenna Specialists Co.

element is enlarged to form a tube. The coaxial transmission line is then fed through the tube, with the center conductor of the cable connected to the radiator, and the shield connected to the tube. The tube acts like the ground plane.

A typical coaxial dipole antenna for the 150- to 174-MHz band has a fairly low vertical radiation angle and a bandwidth of 2 MHz (Fig. 4-26). Heavily built for a power capacity of 1000 watts, the radiating element, the tube or skirt, and a section of supporting pipe have an overall length of about 6 feet.

Adjustable-Pattern and Tower-Mounted Antennas

By using dipoles stacked vertically on a mast or tower, it is possible to take advantage of the natural reflector qualities of the mast or tower. That is, when the spacing between the radiating element and the supporting unit is small, the directional pattern tends to be a cardioid. Farther away from the mast or tower, the stacked dipoles give an omnidirectional pattern. Typically, a stack of two such dipoles provides 3-dB gain with the nondirectional pattern and 6 dB with the cardioid pattern. Six-dB and 9-dB gains are possible with the stacking of four dipoles (Fig. 4-27).

Courtesy The Antenna Specialists Co.

Fig. 4-27. Vertical array of folded dipoles with nondirectional or cardioid radiation patterns for the 150–174-MHz business radio band.

Directional Antennas

For point-to-point systems, selected area coverage, and reduction of backside noise, Yagis, log-periodic, and corner reflector antennas have some use in business radio. Fig. 4-28 shows a five-element Yagi base-station antenna for the 150- to 174-MHz band.

Before selecting one of these types, a careful evaluation of costs and conditions must be made. The directional characteristics and gains of these antennas may not be as advantageous as first examinations might indicate. For example, a base station originally on the outskirts of a city may suddenly find rapid expansion in the area on the backside of its antenna coverage. This would leave the business communications system unable to handle what might be a very important area.

Courtesy Cush Craft Corp.

Fig. 4-28. Five-element Yagi base-station antenna for the 150–174-MHz business radio band.

Even at the vhf and uhf ranges, the corner reflector antennnas are rather large and cumbersome (each side of the reflector for the 150- to 174-MHz range is 4 feet by 5 feet). Vhf Yagi and log-periodic antennas, with 80-inch booms, are also rather cumbersome, but at 406 to 470 MHz these units become easier to handle.

Collinear Antennas

While the collinear antenna array has been known for many years, the newer end-fed and encapsulated types have many advantages. Using a series of half-wavelength radiating elements enclosed within a rigid vinyl or fiber-glass jacket, the vertical collinear has a 5.8-, 7.0-, or 8.26-dB gain and a vertical beam width as small as 8°. Its construction provides for a very low wind resistance, while the vinyl or fiber-glass enclosure protects the radiating elements from the weather (Fig. 4-29).

Although choosing an antenna from among the great variety used for business base stations can be somewhat confusing, geographic and economic factors will usually determine the type selected. In many instances, the initial installation is governed by economy. In the case of temporary building sites—highway and bridge construction, stone quarries, etc.—expensive installations would be wasteful.

Fig. 4-29. Collinear base-station antenna for the 450–470-MHz business radio band. Contains a series of half-wavelength radiating elements enclosed within a fiber-glass jacket.

Courtesy The Antenna Specialists Co.

RELAY STATIONS

Because many two-way systems are on the same carrier frequency, the power is restricted to short-range or local operation. This makes base-to-mobile or mobile-to-base communication extremely difficult when the units are over the horizon from each other or are separated by hills or tall buildings. To overcome this problem, repeater stations are used at some point of high elevation. By talking through the repeater station, base-to-mobile communications over long distances become possible. As long as the repeater antenna can "see" the mobile antenna, two-way communications can be maintained.

If no hills or tall buildings are available, the relay station is sometimes installed as a combined base and relay station. Vehicles maintain their extended mobile-to-mobile coverage as long as they are in "sight" of the base-station antenna. This method of installation is sometimes preferred in small communities, where range requirements are less.

New applications of these repeaters and relay stations are in the uhf frequency bands above 450 MHz. These signals follow extremely straight lines and behave much like rays of light. Coverage halts abruptly when the signals reach the horizon line or encounter hills or other obstacles.

The greatest use of relay stations today is in the various state and city police departments. For state police with a number of posts established about the state, a long-range microwave relay system provides a means of linking all mobile and base stations.

Many large city-police departments now use a highly sophisticated radio system operating in the uhf range, which depends entirely on a number of satellite relay stations strategically located throughout the city. All the officers, both on foot and in patrol cars, carry portable, low-power ("walkie-talkie") units by which, via the various satellite stations, they can communicate not only with the central headquarters dispatcher but also with any other patrolman anywhere in the city. The satellite stations receive the transmissions from the portable units and relay them via telephone lines to the central dispatcher. Transmissions to the officers carrying the portable units are made from a centrally located, high-power transmitter.

5

Mobile Antennas

Although base stations do communicate with other base stations, their primary purpose is to communicate with mobile stations. An exception to this is the amateur radio base station. Many amateurs find working their base stations fully satisfying, while others like the added excitement of mobility.

At the present time, there are approximately 1300 different channels available for the land-mobile service between the frequencies of 25 and 470 MHz. Interference between stations is a serious problem because many users are operating on the same channel at the same time. This situation has been especially serious for business radio users and other services in the 150-MHz band. Amateur radio operators have learned to live with a similar situation on the 10-meter (28-MHz) band, where interference has always plagued the mobile user.

The trend has been to make more use of ultrahigh-frequency (uhf) channels which are less crowded and are capable of reliable communications. Previous objection to the use of uhf equipment has been range limitations. It was felt that fewer miles per transmitter watt would result by shifting to the higher frequencies. The development of new antennas has shown that range can be increased substantially within the uhf spectrum. Users have also found that a relative freedom from impulse-type noise at these frequencies has extended the range limits of the station.

The best known mobile radio services include police, fire, forestry, citizens, amateur, civil defense, business, and marine. Their antenna requirements are basically the same. Mobile antenna systems for use on the 27-MHz Citizens band, for example, are almost identical to those used on the 28-MHz (10-meter) amateur band.

MOBILE ANTENNA TYPES

The basic element of the mobile antenna is the quarter-wave Marconi. Mounted vertically, it is best known as the quarter-wave "whip" and has long been a familiar sight to motorists. A quarter-wave antenna is approximately 8 feet long at 30 MHz, 18 inches at 150 MHz, and a mere 6 inches at 450 MHz.

In its simplest form, the quarter-wave antenna is a vertical wire with the inside conductor of a coaxial transmission line connected to the bottom end. The outside shield of the cable is grounded as in Fig. 5-1. The quarter-wave whip can be mounted on an automobile in one of several ways. The ideal location is in the center of the roof. The whip section is insulated from the rooftop and the coaxial cable connected inside the car, as shown in Fig. 5-2. This type of mounting is practical for the short-length whip antennas at 150 or 450 MHz, but not for the longer whips required for operation on the lower frequencies such as 27 MHz. The 8-foot whips for the 28-MHz amateur band and the slightly longer whips for the 27-MHz Citizens band are mounted at a lower point on the vehicle. Special brackets allow mounting on fenders, bumper, or cowl.

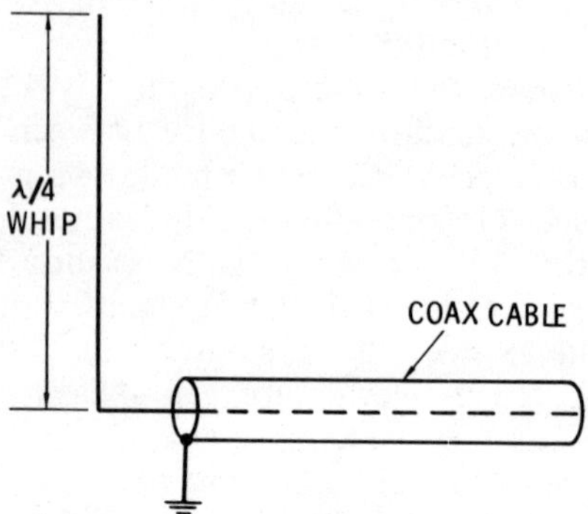

Fig. 5-1. Basic quarter-wave (Marconi) whip antenna.

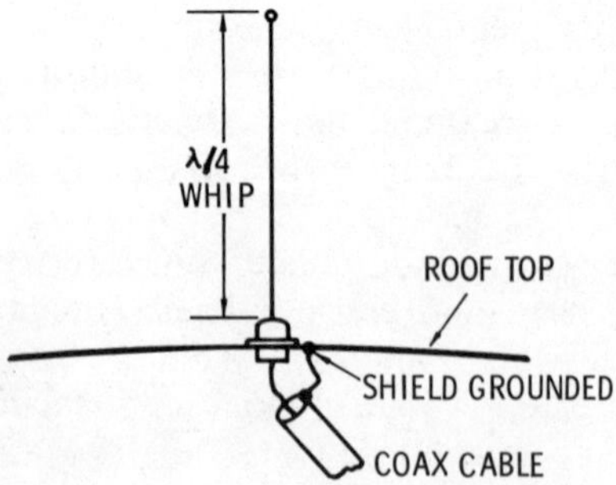

Fig. 5-2. Mounting of quarter-wave whip antenna to automobile rooftop.

ANTENNA LOADING

Possibly the greatest problem encountered with vehicular installations in the 27-MHz region is that of obtaining sufficient ground plane for efficient performance. Installations utilizing bumper mounting result in poor radiation patterns and poor loading/bandwidth characteristics. Another problem develops when the antenna is physically shortened and then does not develop an adequate standing wave and radiation. With new design principles, the 8-foot antenna can be shortened by using the loading-coil principle. The antenna illustrated in Fig. 5-3 is a continuously loaded whip for use on the 27-MHz

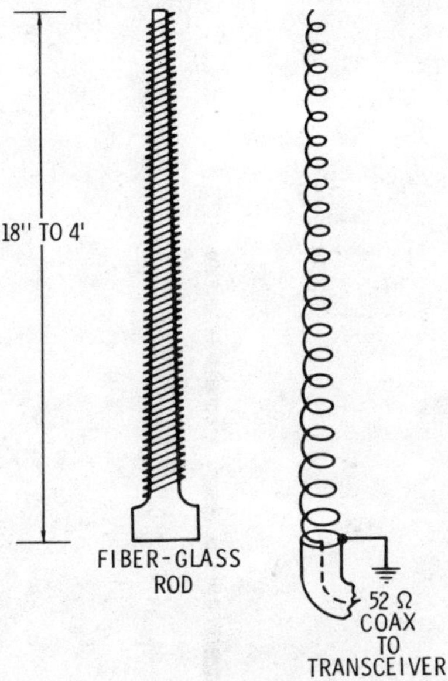

Fig. 5-3. Continuously loaded whip antenna for 27-MHz Citizens band.

Citizens band. A continuously wound step-tapered helical conductor results in a current distribution which is essentially uniform and which produces a 50-ohm match at its resonant frequency. This design principle can be used for antennas having a physical length of from 18 inches to 4 feet. The standard length is 4 feet for most installations, and such antennas can be located on the trunk lid, cowl, or fender.

It is possible to shorten a vertical whip by using a loading coil as a portion of the radiator at its base, as shown in Fig. 5-4. This arrangement was used on the first Citizens-band transceivers. A length of stiff wire is coiled at its base, as in Fig. 5-5, and mounted on an insulator on top of the transceiver. The same principle can be used for hand-held "walkie talkies." Fig. 5-6 shows some typical examples of base-loaded whip antennas for vehicular mounting. The loading coils are

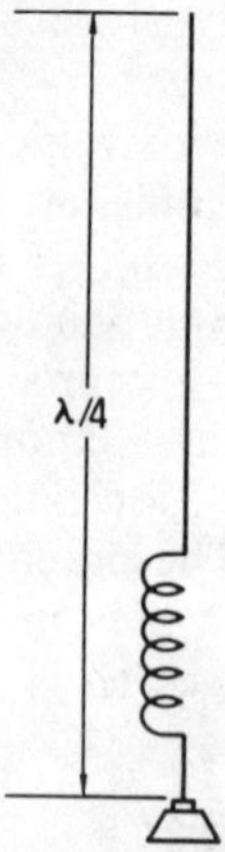

Fig. 5-4. Schematic representation of base-loaded quarter-wave whip antenna.

Courtesy The Antenna Specialists Co.

Fig. 5-5. Base-loaded whip for portable transceivers.

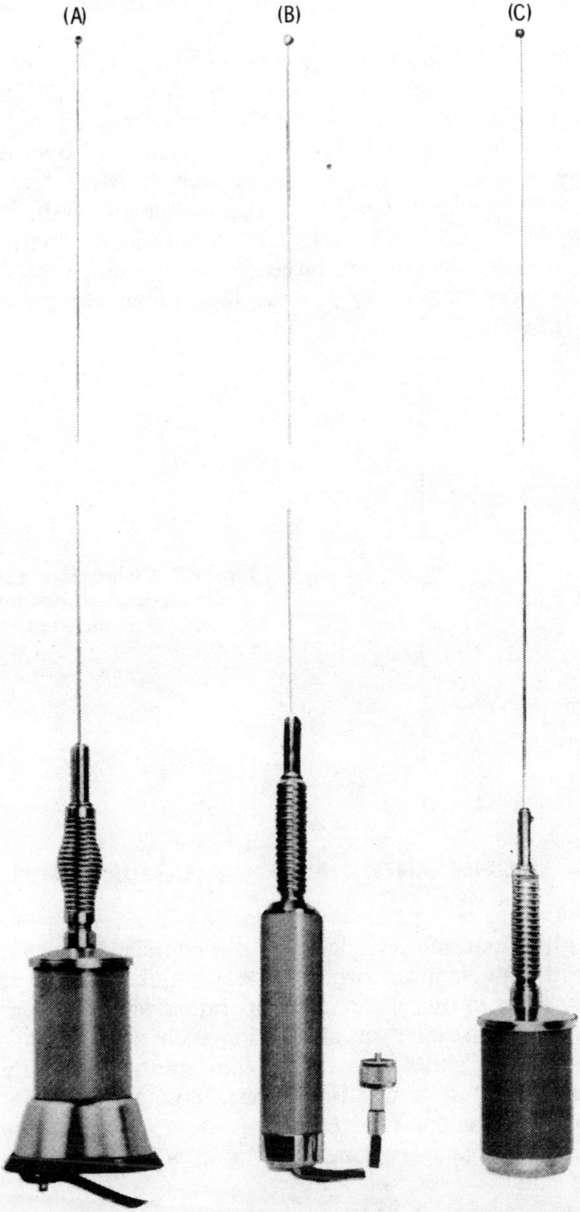

Courtesy The Antenna Specialists Co.

Fig. 5-6. Typical base-loaded CB whip antennas for vehicular mounting. (A) Trunk lid clamp-on mounting. (B) Trunk deck 3/4-inch hole mounting. (C) Rooftop/trunk deck 3/4-inch hole mounting.

in special weatherproof housings just below the stainless-steel shock springs.

The center-loaded whip, as shown in Fig. 5-7, has its loading coil placed near the center of the radiator. Antennas designed for the 27-MHz Citizens band are of the fixed variety. This means that the loading coil and whip combination are required to operate efficiently over only one band of frequencies, in the 27-MHz region. Fig. 5-8 shows some typical examples of center-loaded whip antennas designed especially for large trucks and campers. The dual units at (B) are designed specifically for large diesel rigs and are phased for a broadside array pattern to provide long-range CB communications on the highway.

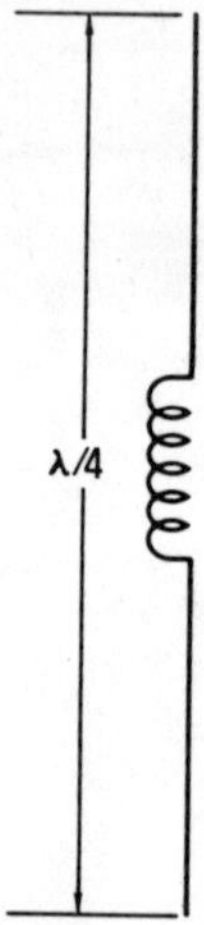

Fig. 5-7. Schematic representation of center-loaded quarter-wave whip antenna.

MOBILE ANTENNAS FOR AMATEUR BANDS

In addition to being a hobby, amateur radio develops certain skills that are often extremely valuable to the public. In times of emergencies—floods, hurricanes, tornadoes, earthquakes, etc.—news reports frequently carry stories of amateur operators providing the only communication with isolated areas. During such times, the amateur displays his greatest skills with mobile/portable radio equipment. Such equipment often has power limitations making it essential to have a good antenna. A good antenna, properly installed and properly matched, can cover several miles, even though the radiated power is extremely small.

The basic mobile antenna used by most amateurs is the quarter-wave vertical, with the vehicle body acting as the ground plane. At the 28-MHz band, the quarter-wave vertical has a length of about 8 feet. It is possible to mount an 8-foot whip-type antenna onto the

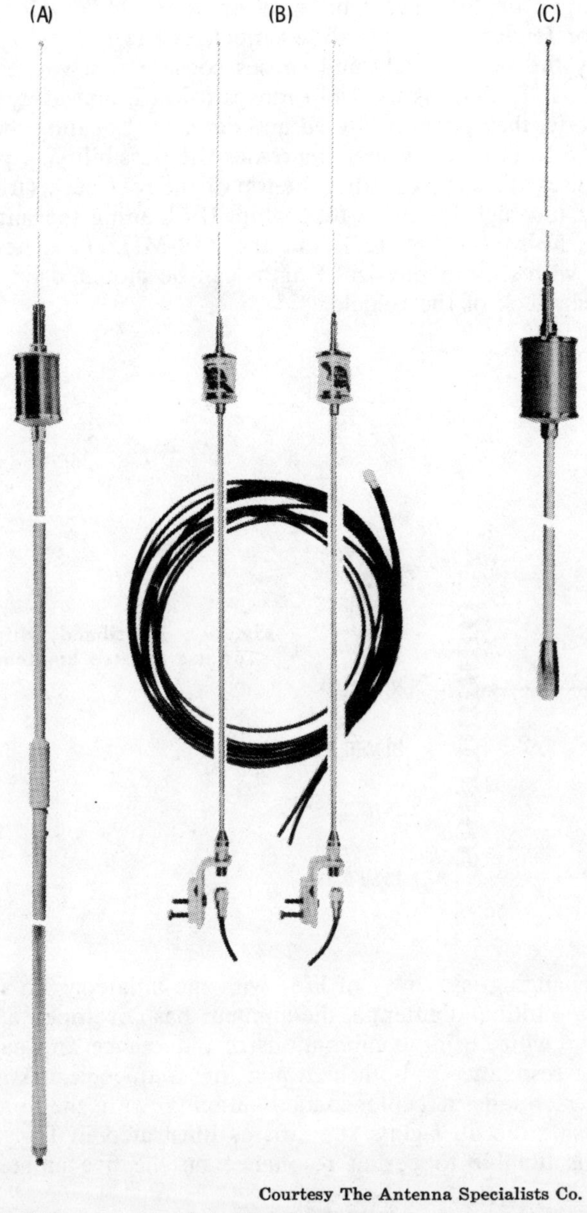

Courtesy The Antenna Specialists Co.

Fig. 5-8. Typical center-loaded CB whip antennas for vehicular mounting. (A) Heavy-duty antenna with fold-down feature for vehicle bumper, deck, side, or camper side mounting. (B) Dual-phased antenna system designed specifically for mirror mounting on large diesel trucks. (C) Heavy-duty antenna designed for swivel-ball mounting on trucks or campers.

rear bumper or the rear fender of an automobile. However, the bumper or fender mounting allows much of the radiation to be absorbed by the vehicle body and causes considerable variation as the whip moves. That is, as the whip moves toward, and away from, the vehicle body, the system is altered and detuned. Locating the antenna toward the front of the vehicle increases the possibility of picking up engine noises. This leaves either the top or the rear deck (trunk), and the top is too high for the 8-foot whip. By loading the antenna and using the 50-MHz (6-meter) and the 144-MHz (2-meter) bands, antennas with shorter physical lengths can be mounted on either the top or rear deck of the vehicle.

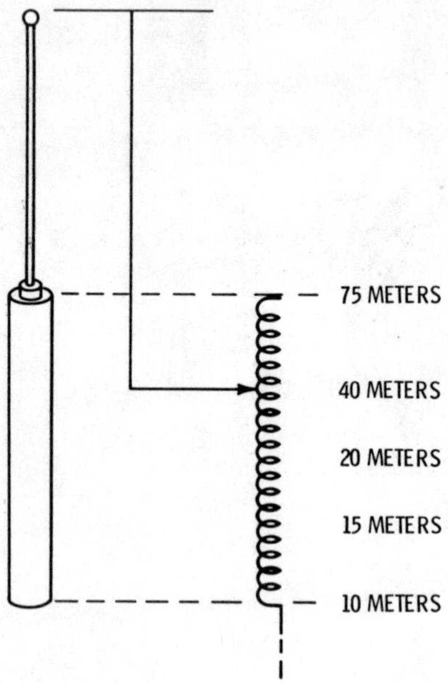

Fig. 5-9. Multiband whip antenna for use on five amateur bands.

Band changing is a "way of life" with the amateur. To satisfy the need for a multiband antenna, the amateur has developed a means of loading the whip, using combinations of inductance and capacitance to provide resonance at both high and low frequencies. A multiband whip is essentially a center-loaded antenna with the loading coil wound inside the fiber-glass support, as illustrated in Fig. 5-9. This inductor is tunable to permit resonance on the five amateur bands shown.

Insulation has been removed from a portion of each coil turn so that the coil is internally exposed to a circular contactor connected to the top whip section. The whip is constructed so that it can be moved in and out of the loading coil section. A locking feature permits the whip to be secured at any frequency setting throughout the five bands, and is marked to show the setting for each band. This type of loading permits exact antenna resonance to be obtained anywhere within a given band. The overall length of the antenna with the whip fully extended is 9 feet, 9 inches. The height of the support column, including the loading coil section, is 63 inches at its minimum length (whip fully depressed inside the loading section). Another version of this multiband whip has the 10-meter (28-MHz) position at the top of the inductor and lower positions for the 6- and 2-meter bands.

The half-wave or ⅝-wavelength antenna, having a large portion of its standing wave on its upper end, has a greater ground wave. Its radiation is also farther away from the body of the vehicle and therefore is less affected by the body. This contrasts with the quarter-wavelength vertical that radiates most of its energy from its lower end, where the vehicle body absorbs some of the radiation. It is possible to load the 8-foot whip and develop an antenna having an electrical length of ⅝ wavelength at the 28-MHz frequency. This, of course, is much easier at 50 MHz where the electrical ⅝ wavelength is about 12 feet. At 144 MHz, the electrical ⅝ wavelength is only slightly more than 50 inches, and this size antenna can be readily mounted on the rear deck of a vehicle.

Matching the antenna impedance to that of the transmission line and transceiver is highly important in achieving maximum radiation. While the vertical antenna is assumed to have an impedance of 50 ohms, its mounting on a vehicle can alter that factor considerably. CB and business radio often have ample power and radiation without being overly concerned with impedance matching. The amateur, however, is never content with his radiation being "just enough"—he really never knows when it may be necessary to reach a bit farther with his signals. For this reason, the amateur uses every possible means to obtain an impedance match. On the mobile unit, the quarter-wavelength matching stub cannot be used, as it is at the base station, because of its size. Fig. 5-10 shows an "L" network used for matching the line to the antenna. With a loaded antenna, the inductor L_m becomes a portion of the loading coil. Capacitor C_m is then adjusted to acquire a minimum swr (standing-wave ratio). This assumes that the antenna and its loading coil have been made to resonate at the desired frequency and will emit maximum radiation.

Antenna construction has been somewhat revolutionized by the use of vinyl tubing and fiber-glass rods. Such tubing and rods can be cut, drilled and machined to almost any form and still provide flexibility and strength. In addition to the mobile antennas already discussed, vinyl tubing and fiber-glass rods are also used in constructing mobile antennas for camping or temporary base stations. Such portable antennas can be taken apart in sections and then reassembled in a short time. Fig. 5-11 shows one type of fiber-glass antenna that uses a temporary "no-hole" trunk mounting. This antenna also has an adjustable tip for precise tuning.

Still another factor to be considered with mobile antennas for amateur bands is the reduction of noise produced by engine ignition. The

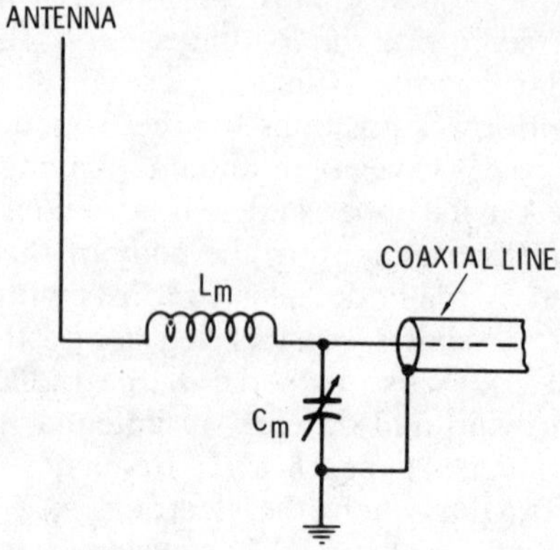

Fig. 5-10. An "L" matching network.

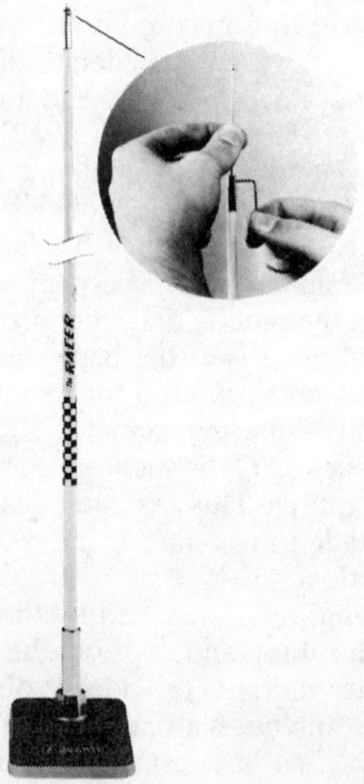

Courtesy Avanti Research and Development, Inc.

Fig. 5-11. Fiber-glass mobile antenna with adjustable tip for precise tuning.

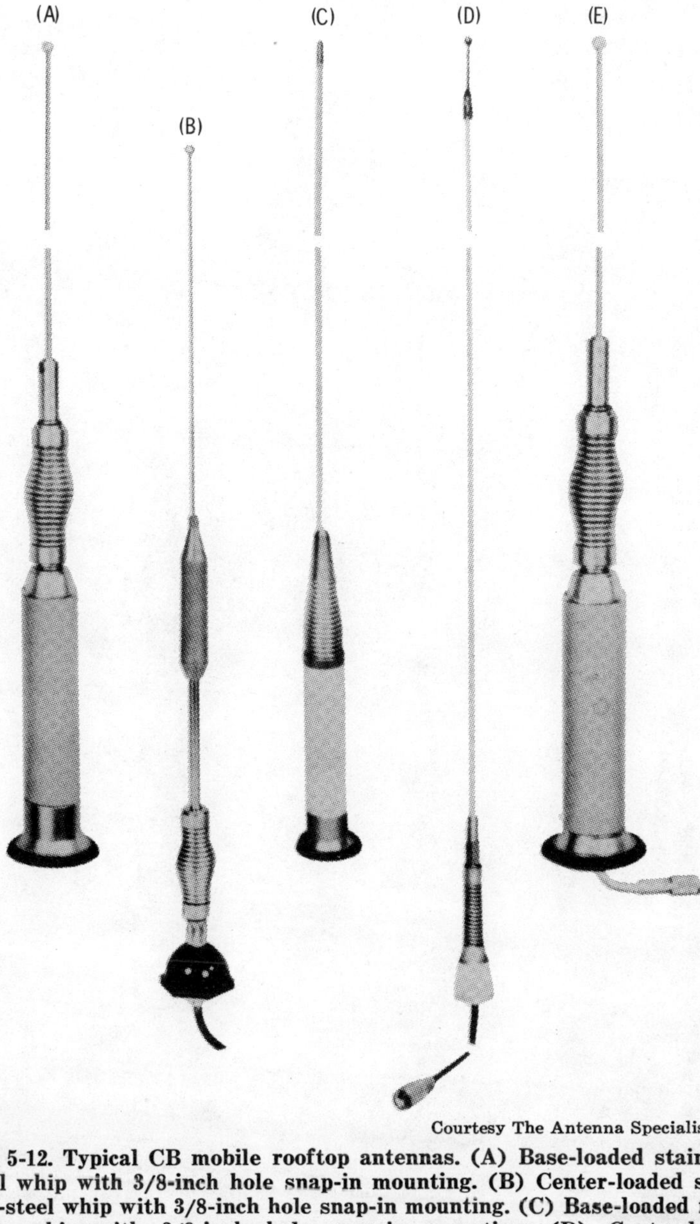

Courtesy The Antenna Specialists Co.

Fig. 5-12. Typical CB mobile rooftop antennas. (A) Base-loaded stainless-steel whip with 3/8-inch hole snap-in mounting. (B) Center-loaded stainless-steel whip with 3/8-inch hole snap-in mounting. (C) Base-loaded fiber-glass whip with 3/8-inch hole snap-in mounting. (D) Center-loaded fiber-glass whip (winding inside fiber glass) with tuneable tip and 3/8-inch hole mounting. (E) Base-loaded stainless-steel whip with 3/4-inch hole snap-in mounting.

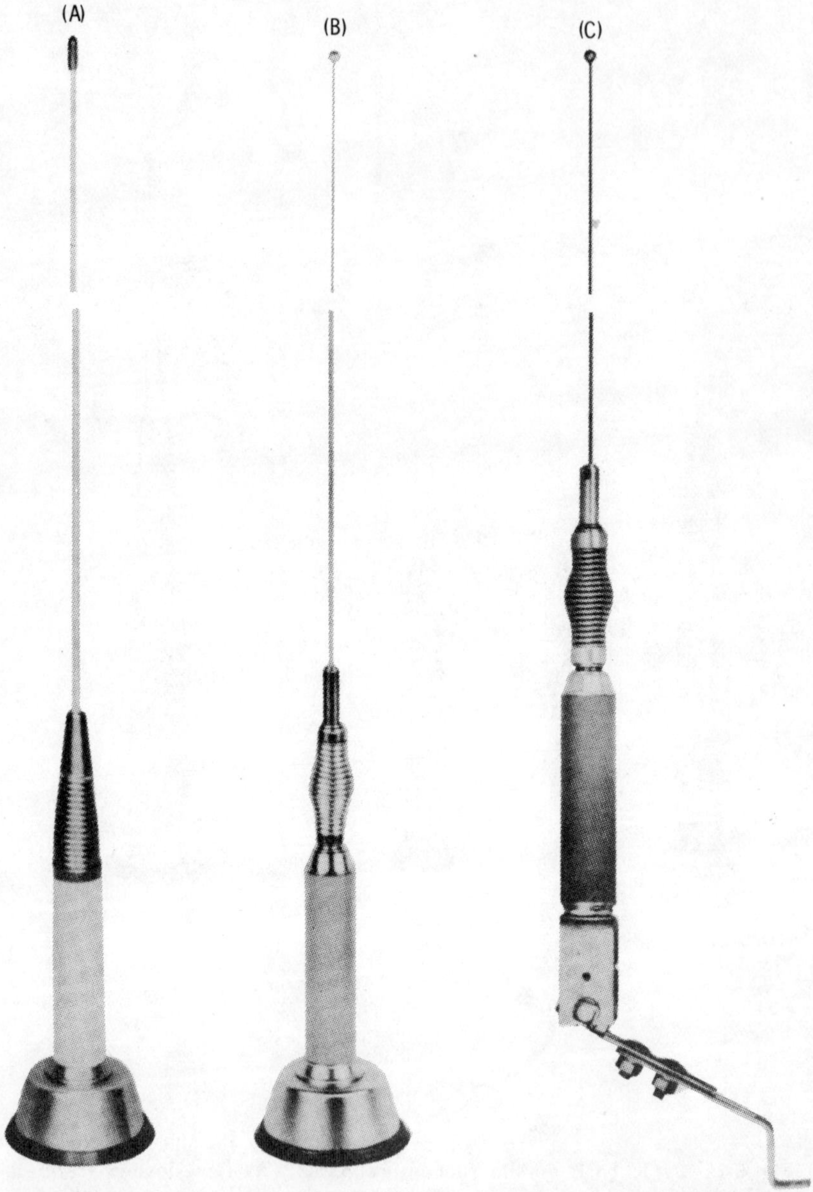

Courtesy The Antenna Specialists Co.

Fig. 5-13. Typical CB mobile trunk-mount antennas. (A) Base-loaded fiberglass whip with trunk lid clamp-on mounting. (B) Base-loaded stainless-steel whip with trunk lid clamp-on mounting. (C) Base-loaded stainless-steel whip with trunk groove mounting.

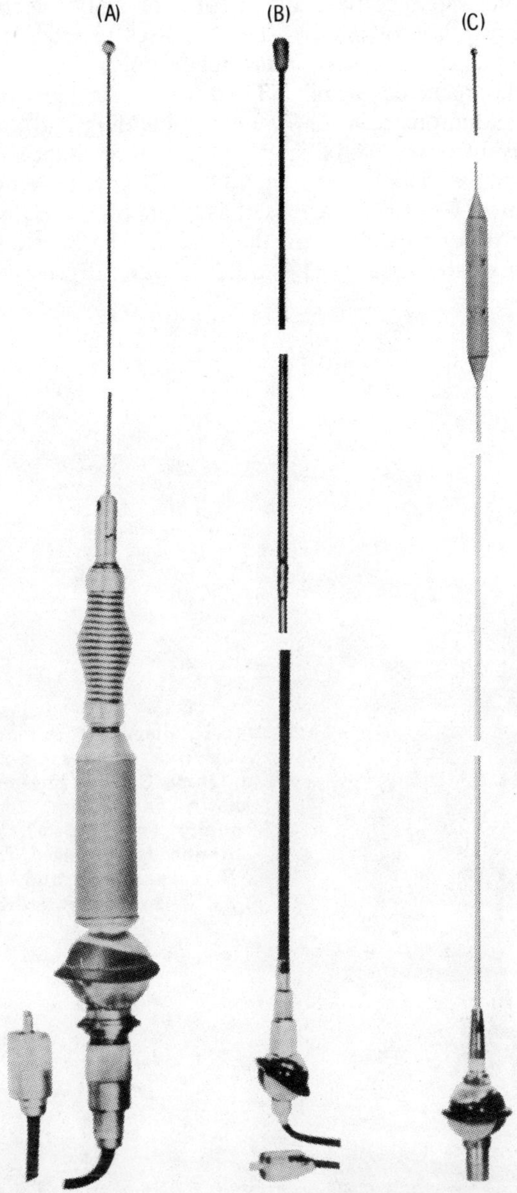

Courtesy The Antenna Specialists Co.

Fig. 5-14. Typical CB mobile cowl-mount antennas. (A) Base-loaded stainless-steel whip. (B) Center-loaded fiber-glass whip with loading coil wound into fiber glass. (C) Center-loaded stainless-steel whip for CB and a-m/fm radio.

engine ignition system is basically a form of static electricity which develops sharp pulses of energy. These pulses produce radiations at nearly all frequencies, causing interference noises on both the transmitted and the received signals. Therefore, it is very important to reduce these radiations to a minimum by shielding and grounding. In addition to using resistor-type spark plugs or resistance-wire cabling to the spark plugs, it is best to shield each spark plug and its cable to the distributor. Capacitors are used as filters on the leads to the generator or alternator as well as on the leads of the heat and fuel-level gauges. Some of the areas that should be carefully bonded together

Fig. 5-15. Dual-purpose stainless-steel "disguise" antenna made to look like ordinary a-m car radio antenna. Coaxial line-matching stub shown at bottom, together with coupler (not shown), permits this antenna to be used for both the CB transceiver and the a-m/fm broadcast receiver.

Courtesy The Antenna Specialists Co.

electrically (grounded) include the engine, air cleaner, exhaust lines, battery ground terminal, steering column, hood, front and rear bumpers, and trunk lid. Wheels and tires also tend to build up static electricity, while the antenna itself, insulated from the rest of the body, often acquires a static charge just before or during an electrical storm.

MOBILE ANTENNAS FOR CB

Operating on a narrow frequency band from 26.965 to 27.255 MHz, Class D Citizens band radio uses a single antenna covering all of its 23 channels. Unlike the amateur mobile antenna, this antenna has no need of band switching, and its power level is much lower. This lower power level limits the CB radiation coverage. However, coverage can be extended appreciably with the proper base and mobile antenna combination. That is, both the base antenna and the mobile antenna for a CB system play a part in extending the coverage. For this reason, the CB mobile antenna must be selected and installed correctly.

The basic CB mobile antenna is the loaded quarter-wave whip which uses the vehicle body as its ground plane. With the different mountings, shock springs, and loading coils, these quarter-wave whips range from 18 to 46 inches in height. For mounting on the roof of the vehicle, the units use a 3/8- or 3/4-inch hole and snap in with a plastic ring or are secured with a locknut (Fig. 5-12). Other types clamp to the front side of the trunk lid or bolt into the outer groove under the trunk lid (Fig. 5-13). Cowl-mounting types (Fig. 5-14) are of the single-hole, locknut variety having a swivel arrangement that permits adjustment to the surface angle. The "disguise" antenna, an innovation copied from police radio, uses the same whip for the CB transceiver and the a-m/fm broadcast receiver (Fig. 5-15). Such a dual-purpose antenna will not perform as well as a single antenna tuned to a specific frequency; however, the dual-purpose disguise antenna is likely to be satisfactory in most situations. Disguise antennas use coaxial line-matching stubs and have a coupler separating the CB and the a-m/fm signals.

For special purposes, such as recreational vehicles, boats, and trucks, there are some half-wave and 5/8-wave antennas. Most of these half-wave and 5/8-wave antennas are center loaded; however, a few are a full half wavelength (18 feet) for use on boats. Half-wave and 5/8-wave antennas have the advantage of not requiring a ground plane as well as having better vertical radiation patterns. Many of these longer antennas are vinyl or fiber-glass protected and can be folded down or taken apart.

The big-rig diesel truckers are now making extensive use of directional antenna arrays for CB communications. By taking advantage of the truck's width of nearly 9 feet (¼ wavelength), it is possible to develop a form of the broadside array (Fig. 5-16). These two antennas are fed in phase and provide an elongated radiation pattern covering the forward, rear, and side portions of the highway of immediate interest to the trucker. Some antenna manufacturers offer a mobile "scanner" for use with phased antennas mounted a quarter-wavelength apart, as on the truck. These devices vary the phasing of the power fed to the two antennas, producing varying radiation patterns. In effect, the scanner changes the direction from which the radio waves are received or sent.

MOBILE ANTENNAS FOR BUSINESS RADIO

Mobile antennas play a more important role in business radio than they do in either CB or amateur radio. In most instances, the primary

Fig. 5-16. Center-loaded dual-phased CB antenna system mounted on a large diesel truck.

reason for a business radio system is communication with mobile units. And yet the mobile business unit commonly has less need for an expensive or elaborate antenna.

As with the other services, the basic mobile antenna for business radio is the quarter-wavelength vertical whip requiring a ground plane. The body of most vehicles provides sufficient ground plane (or ground plate as it is sometimes known). The vhf business radio band (between 150 and 174 MHz) quarter-wavelength whip has a maximum length of 18¾ inches. This short length allows the quarter-wavelength whip to be mounted on either the roof or the trunk deck; the latter being preferred because of possible low clearances. On the 406- to 470-MHz business radio bands, the quarter-wavelength whip maximum length is less than 7 inches, and its mounting on the roof

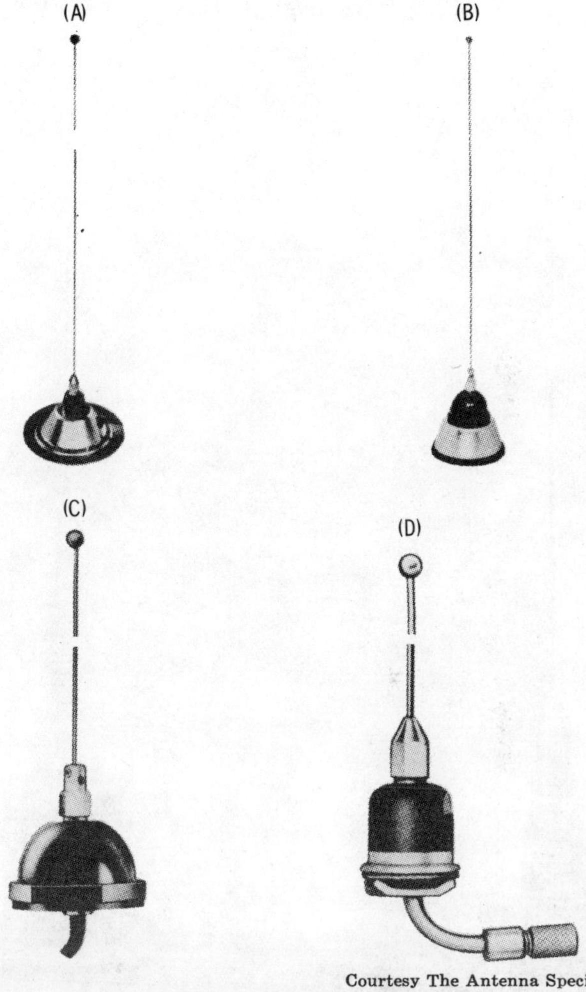

Courtesy The Antenna Specialists Co.

Fig. 5-17. Typical quarter-wave whip antennas for business radio. (A) Magnetic mount for temporary installation on rooftop, trunk deck, or any flat surface. (B) Rooftop or trunk deck 3/8-inch hole snap-in mount. (C) Heavy-duty rooftop or trunk deck, 3/4-inch hole mount with toggle-type mounting plate. (D) No-hole, trunk lid clamp-on mount.

of the vehicle is simple. Rooftop mounting is usually accomplished by means of a ⅜- or ¾-inch hole directly above the interior dome light. Removal of the dome light provides access to the underside of the roof without otherwise damaging the interior. Most rooftop mountings are of the locknut variety, but there are snap-in types as well as magnetic types; the magnetic types are for temporary use but are capable of staying on at legal speed limits. For trunk mounting, setscrews provide the clamping force over the upper edge of the trunk lid.

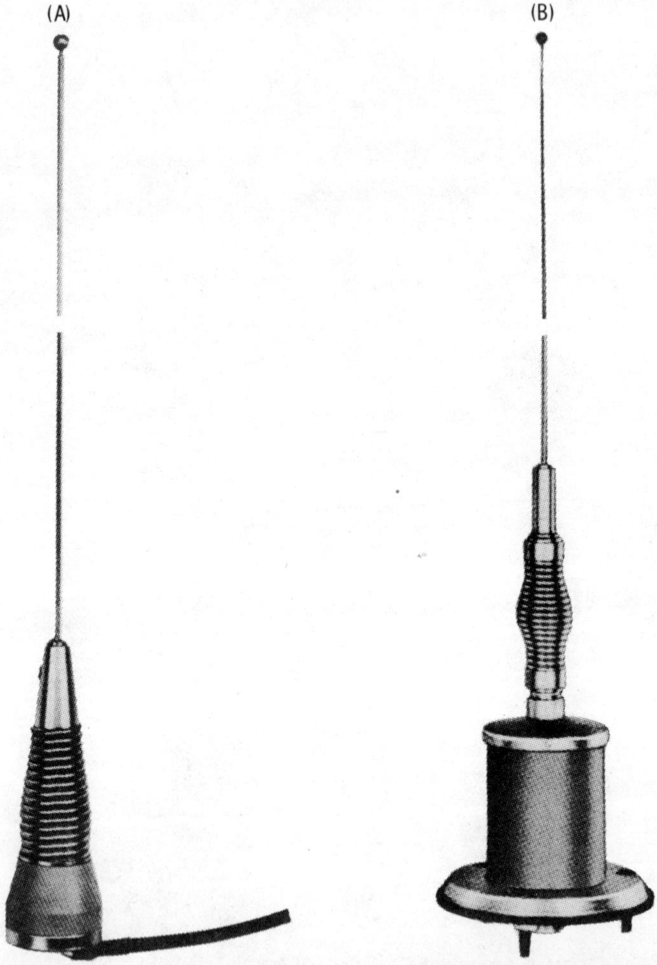

Courtesy The Antenna Specialists Co.

Fig. 5-18. Base-loaded, 3-dB gain, 5/8-wavelength mobile antennas for the 150- to 174-MHz business radio band. (A) Rooftop or trunk deck automobile mount. (B) Heavy-duty truck and other diesel vehicle mount.

Operationally, the quarter-wavelength whip provides an omnidirectional radiation pattern with an impedance of 50 ohms. Fig. 5-17 shows some typical quarter-wave whip antennas for business radio.

To achieve a 3-dB gain, 49-inch whips are base-loaded, giving an effective ⅝ wavelength in the 150- to 174-MHz range (Fig. 5-18). At the higher 450- to 470-MHz range, two ⅝-wavelength radiators are stacked in a collinear array to give a 5-dB gain (Fig. 5-19). The

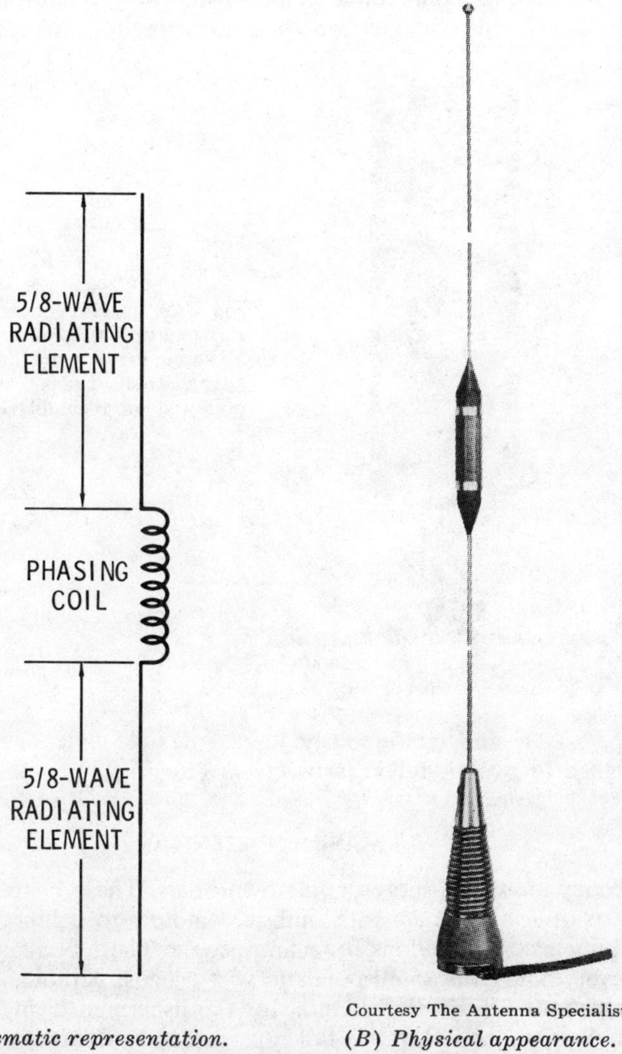

Courtesy The Antenna Specialists Co.

(*A*) *Schematic representation.* (*B*) *Physical appearance.*

Fig. 5-19. Collinear, phased 5/8-wavelength mobile antenna system for the 450- to 470-MHz business radio band.

⅝-wavelength antenna concentrates the vertical radiation to a beam width of about 20°, while the collinear array has a vertical beam width of 12° or less.

In addition to these basic antennas used for mobile business radio, there are a number of special-purpose types, such as the "disguise" antennas used by police, concentric cylinder types used by railroads (Fig. 5-20), fiber-glass whips for motorcycles, coaxial antennas for vehicles having nonmetallic bodies with no ground-plane surface (Fig. 5-21), low-silhouette antennas enclosed within plastic hoods

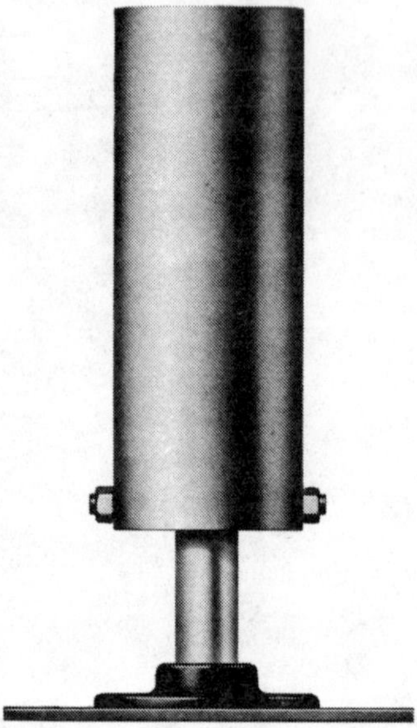

Courtesy The Antenna Specialists Co.

Fig. 5-20. Special concentric-cylinder railroad antenna for the 150- to 174-MHz business radio band. Made of silverplated brass and can be mounted on locomotive or caboose.

(Fig. 5-22), and continuously loaded flexible antennas specifically designed for walkie-talkie transceivers (Fig. 5-23).

MARINE ANTENNAS

Today more and more people own boats. These boats range from canoes or fishing boats with outboard motors to cabin cruisers with sleeping accommodations for many people. These boats are operated on everything from small ponds to vast oceans. Although radio communications are hardly essential for the fisherman trying his luck in a small pond, the CB walkie-talkie is frequently a part of a fisherman's equipment. The commercial or private boat operating on the high seas or larger inland lakes finds radio-communication to be essential.

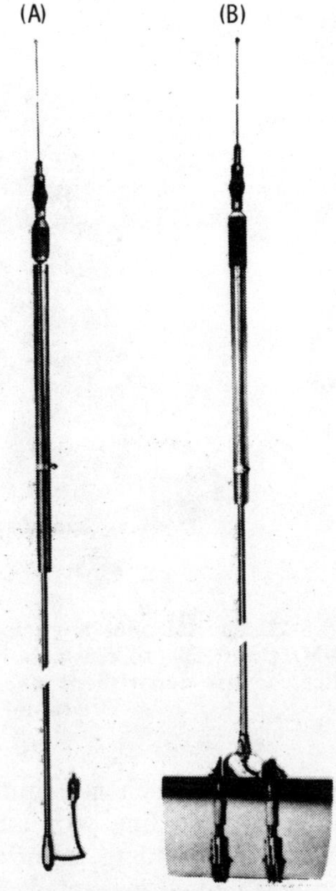

Courtesy The Antenna Specialists Co.

Fig. 5-21. Special coaxial antennas for the 150- to 174-MHz business radio band. These antennas are particularly well-suited for use on vehicles having nonmetallic bodies. (A) Swivel-base mounting. (B) Double chain link, rear-bumper mounting.

Most saltwater areas, including bays and backwaters, are under the jurisdiction of the Coast Guard and radios are mandatory. The Coast Guard also controls much of the activity on the Great Lakes and the navigable rivers.

Freighters, tankers, and other sea-going vessels that can be referred to as ships, rather than boats, use frequencies and antennas that are beyond the scope of this book. Such ships move in international waters where their communications are controlled by international agreement. The boats that we are concerned with use both the Class D Citizens band and a 156- to 163-MHz band specifically assigned to marine use. On the CB channels (26.965 to 27.255 MHz), the most popular antenna for boats is the half-wavelength vertical made of fiber glass or stainless steel. Such half-wave antennas require no ground plate, but there should be adequate grounding by means of the water. Since many boats are now being built of fiber glass, it is

Courtesy The Antenna Specialists Co.

Fig. 5-22. Special, plastic-enclosed, low-silhouette antenna for the 150- to 174-MHz and 450- to 470-MHz bands. Designed especially for railroad, bus, police, or fire-department applications. This antenna has an omnidirectional radiation pattern.

necessary to attach aluminum foil to the underside of the boat for adequate grounding. An adhesive-backed foil is available for this purpose. Depending somewhat on operational conditions, most marine antennas have few restrictions on height and use full half-wavelength whips of 18 feet or a loaded half-wavelength of 4 to 8½

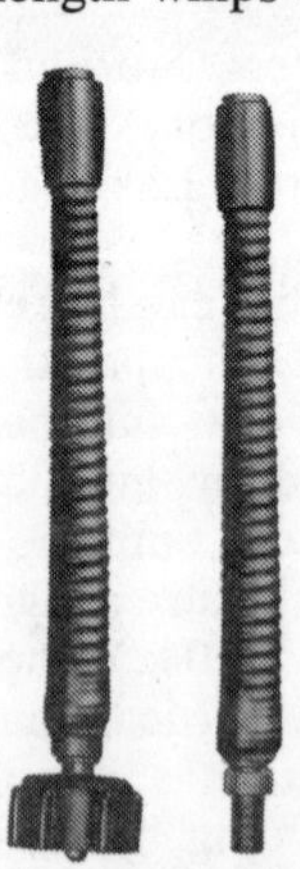

Courtesy The Antenna Specialists Co.

Fig. 5-23. Special, continuously loaded, flexible whip antennas designed for portable and walkie-talkie applications on the 150- to 174-MHz and 450- to 470-MHz bands. Two different mounting styles are shown for fitting to different manufacturers' transceivers.

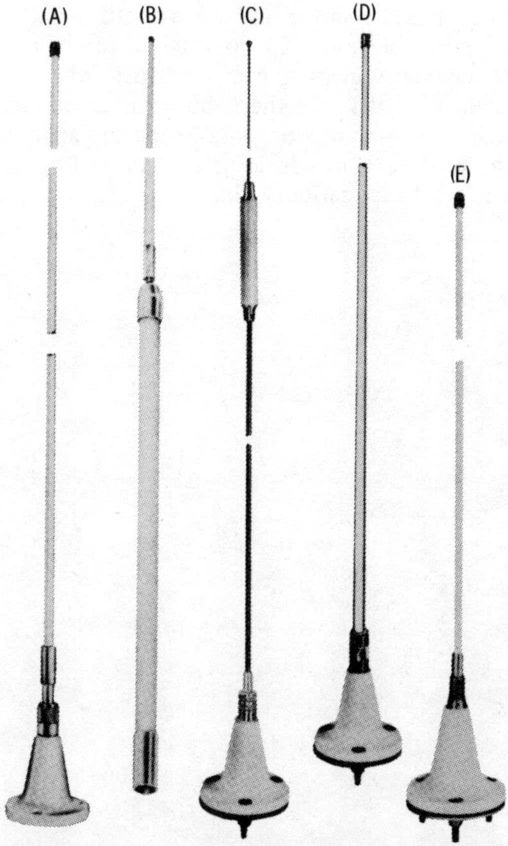

Courtesy The Antenna Specialists Co.

Fig. 5-24. Typical loaded and unloaded half-wavelength antennas for marine radio use. (A) Unloaded (full half-wavelength) fiber-glass whip with 3-hole mounting and fold-down feature. (B) Unloaded (full half-wavelength) fiber-glass whip with female fitting for attaching to standard marine mount. Two-piece construction for easy take-down and storage. (C) Center-loaded stainless-steel whip with 3-hole mounting and fold-down feature. (D) Center-loaded fiber-glass whip (loading coil wound into fiber glass) with 3-hole mounting and fold-down feature. (E) Center-loaded fiber-glass whip (loading coil wound into fiber glass) with 3-hole mounting and screw-thread fitting for removing whip.

feet. Many of these have a fold-down or take-apart arrangement. Fig. 5-24 shows some typical antennas designed for marine use.

Depending somewhat on the distance to be covered, antennas for the 156- to 163-MHz band can be of the common quarter-wavelength type up to a collinear arrangement of ⅝-wavelength elements. The quarter-wavelength type with its need for a ground plane becomes rather awkward, and the single half-wavelength or 5/8-wavelength antenna lacks appreciable gain. In contrast, a collinear array of two 5/8-wavelength elements gives a gain of 6 dB, while a four-element array has a gain of 9 dB. On the cabin cruiser or larger boat, it is possible to arrange these antennas in a broadside array giving a directional pattern. With this broadside array and its large gain, the boat operator has a good navigational aid.

Index

A

Abnormal propagation, 16-18
Adjustable pattern antenna, 93
Altitude effect on radio wave velocity, 17
Amateur antennas, 73-83
  mobile, 104-110
Amplitude, 20-21
Angle of reflection, radio waves, 10
Antenna(s)
  adjustable pattern, 93
  amateur, 73-83
    mobile, 104-110
  business radio, 90-95
  Citizens band, 84-90
    mobile, 113-114
  coaxial dipole, 92-93
  collinear, 95
  directional
    business radio, 94-95
  directivity, 32
  fundamentals, 25-44
  gain, 10, 41-43
  ground plane, 48, 84-86
    business radio, 92
  ground system, 81, 83
  half-wave, 36, 45-47
  Hertz, 74-75, 79-80

Antenna(s)—cont
  horizontal, 45-47
    amateur, 74-78
  horizontal/vertical, 89-90
  length, 36-38
  loading, 101-104
  log periodic, 68-71
  long wire, 65-67
  Marconi, 78
  marine, 118-122
  microwave, 62-64, 71-72
  mobile types, 100
  parabolic, 63-64
  patterns, 31-33
  pattern development, 33-35
  rhombic, 66-67
  sensitivity, 38
  unidirectional, 38
  vertical, 47-50
    amateur, 78-83
    Citizens band, 84-89
  very low frequency, 65
  Yagi, 67-68, 87-88
  Zepp, 75-76
Array(s), 50-56
  broadside, 52-53
  collinear, 53-54
  end fire, 51-52
  feeding of, 54-56

Array(s)—cont
  stacked, 59-60
Atmospheric pressure, 16-17

B

Base loaded antenna, 101-103
Basic antenna characteristics, 29-31
Beam-width angle, 38-39
Bow-tie antenna, 61-62
Broadside array, 52-53
Business radio antennas, 90-95
  mobile, 114-118

C

Center fed antenna, 75-76
Center loaded antenna, 104
Characteristic impedance, 26, 27, 28
Circular polarization, 20
Citizens band antennas, 84-90
  mobile, 113-114
Coaxial dipole antenna, 92-93
Collinear
  antenna, 95
  array, 53-54
Communication satellites, 15-16
Constants, space, 35-36
Constant-voltage/constant-current device, 28
Corner reflector, 61
Cosmic rays, 7, 11
Counterpoise, 79
Current-fed doublet, 76
Curved radio signals, 16, 17

D

Delta feed, 68
Dielectric constant, space, 35
Dipole, 45
  antenna, 31
  fan, 25
  half-wave, 25
    folded, 25
  horizontal, 38
  simple, 25
  standard, 42
Directional antenna
  business radio, 94-95
Directivity
  and gain, 38-44
  antenna, 32
Director, 58-59
Disguise antennas, 113
D layer, 11
Doublet, half wave, 76
Duct, 17-18

E

Effective radiated power, 43
E layer, 12
Electric field
  component, 18-19, 20
  vector, 19-20
Electromagnetic radiation, 7-8
Electromotive force (emf), 20
Elements, parasitic, 56-59
End-fire array, 51-52
Evolution of antenna, 30

F

F_1, F_2 layers, 12
Feed, antenna array, 54-56
Field strength, 20-21
Folded dipole, 45-47
Four-terminal network, 26, 27, 28
Free electrons, 11, 12, 15
Frequency
  and wavelength, 35-36
  bands, 21-23
  calculation of, 36
Full-wave length antenna, 32-33

G

Gain
  antenna, 41-43
  receiving, 41
  transmitting, 41
Gamma match, 87-88
Generator, 26-27
Graph, polar, 38

Ground
marine antennas, 119-120
plane antennas, 48, 84-86
business radio, 92
system, amateur antenna, 81, 83
-wave attenuation, 9-10
waves, 8

H

Half-power point, 38
Half-wave antenna, 31, 36, 45-47
Half wavelength, 36
Heat waves, 7
Hertz antenna, 74-75, 79-80
High frequency (hf), 21
wave, 8
Horizontal
antenna, 45-47
amateur, 74-78
dipole, 38
pattern, 31
Horizontally polarized wave, 19-20
Horizontal/vertical antenna, 89-90

I

Ignition noise, reduction of, 112-113
Impedance matching, delta, 68
Incident wave, 29
Index of refraction, 16, 17
Ionized particles, 10
layers of, 11-12
Ionosphere, 8, 10-12
dense, 14-15
effect on radio waves, 12-16
reflection from, 18
Ionospheric refraction, 12
Ions, 11

L

Lamda (λ), 34
Laws of refraction, 13
Length, antenna, 36-38
Lightning protection, 83
Light ray travel, 13-14
Light waves, 7
L network, 107-108
Loading, 50, 101-104
Lobes, 38-40
Log periodic antenna, 68-71
Long wavelength, 8
Long wire antenna, 65-67
Low frequency (lf), 21
wave, 8

M

Magnetic field
component, 18-19, 20
vector, 19
Marconi antenna, 47, 78
Marine antennas, 118-122
Mathematical relationship, wave velocity, frequency, and wavelength, 8
Medium frequency (mf), 21-22
Microwave antennas, 62-64, 71-72
Mobile antennas
amateur, 104-110
business radio, 114-118
Citizens band, 113-114
types of, 100

N

Negative ions, 11
Nighttime transmission, 12
Nodes, standing wave, 29
Nulls, 40

O

Oersteds, 20
Omnidirectional pattern, 47

P

Parabola, 64
Parabolic
antenna, 63-64
reflector, 62-64
Parasitic elements, 56-59
Particles, charged or ionized, 10
Passive satellites, 18
Pattern(s)
adjustable, 93

Pattern(s)—cont
  directivity, 38
  horizontal, 31
  vertical, 31
Permeability constant, space, 35
Polar
  diagram, 31
  graph, 38
Polarization, 18-20
Positive ions, 11
Power, effective radiated, 43
Propagation, sky-wave, 10

Q

Quarter-wave antenna, 47-48

R

Radiation(s)
  antenna, 10
  electromagnetic, 7
  of energy, 30-31
  pattern, determining, 35
Radio wave(s), 7-10
  and wave propagation, 7-23
  components, 18-19
  reflection, 15
  velocity, 16-17
    effect of
      altitude, 17
      pressure, 16-17
      temperature, 16-17
      water vapor, 16-17
Receiving gain, 41
Reflector(s), 57-58, 61-64
  corner, 61
  parabolic, 62-64
Refraction
  basic characteristic of, 14
  complete, 15
  index of, 17
  ionospheric, 13-14
Relay stations, 96-97
Repeaters, 96
Response pattern, effect of directors on, 59
Rhombic antenna, 66-67

S

Satellites
  communication, 15-16
  passive, 18
Sensitivity, antenna, 38
Short wavelength, 8
Skip
  distance, 13
  transmission, 12
Sky wave(s), 8
  propagation, 10
  reflection, 11
Soil resistance, 9
Space constants, 35-36
Speed of light, 8
Stacked arrays, 59-60
Standard dipole, 42
Standing wave, 28, 29, 30
  nodes, 29
  ratio (swr), 29
Static electricity, 113
Superhigh frequency (shf), 22-23

T

Temperature
  effect on wave propagation, 16-17
  inversion, 17-18
T-match, 88-89
Transmission line
  characteristics, 26
  principles, 25-29
  reflection, 28-29
Transmission medium, 8
Transmitting antennas, radiation of, 10
Transmitting gain, 41
Troposphere, 8
Tropospheric scatter propagation, 18
Turnstile arrangement, 47
Two-way mobile radio communications, 34

U

Ultrahigh frequency (uhf), 22
Ultraviolet radiation, 11, 12
Unidirectional antenna, 38

V

Vee antenna, 65-66
Velocity
  of propagation
    factor, 54-55
    metallic conductor, 38
  of radio waves, 16-17
  wave, 7
Vertical
  antenna, 31-32, 47-50
    amateur, 78-83
    Citizens band, 84-89
  pattern, 31
  polarized wave, 19-20
Very high frequency (vhf), 22
Very low frequency (vlf), 21
  antenna, 65

W

Water vapor, 16-17
Wave(s)
  ground, 8, 9
  propagation, 7-23
  radiation, electromagnetic
    amplitude, 7
    direction, 7
    frequency, 7, 8
    polarization, 7
    velocity, 7, 8
Wave(s)—cont
  radio, 7-10
  sky, 8, 9
  transmission medium, 7
    air, 7
    copper, 7
  velocity
    changes, 7
    frequency and wavelength, mathematical relationship of, 8
Wavefront, 9
Waveguide, 71
Wavelength, 28
  calculation of, 35-36
Whip antenna
  base loaded, 101-103
  center loaded, 104
  continuously loaded, 101

X

X-rays, 7

Y

Yagi antenna, 67-68, 77-78, 87-88, 94-95
Y feed, 68

Z

Zepp antenna, 75-76